Advances in Experimental Medicine and Biology

Volume 938

More information about this series at http://www.springer.com/series/5584

Miriam Ramírez-Domínguez
Editor

Pancreatic Islet Isolation

From the Mouse to the Clinic

Springer

Editor
Miriam Ramírez-Domínguez
Laboratory of Cell Therapy of Diabetes,
Department of Pediatrics, Faculty
of Medicine and Odontology, Hospital
Cruces
University of the Basque Country
(UPV/EHU)
Leioa, Biscay, Spain

ISSN 0065-2598 ISSN 2214-8019 (electronic)
Advances in Experimental Medicine and Biology
ISBN 978-3-319-39822-8 ISBN 978-3-319-39824-2 (eBook)
DOI 10.1007/978-3-319-39824-2

Library of Congress Control Number: 2016948757

Printed on acid-free paper

This Springer imprint is published by Springer Nature
The registered company is Springer International Publishing AG Switzerland

I dedicate this book to my family, for their unconditional love and support.

Preface

Pancreatic islets contain the beta cells, which are the source of insulin in the body, and hence, in terms of diabetes, they are fundamental structures for any preclinical in vitro or in vivo studies in animal models or related to human transplantation. However, human and animal islet isolation has been described as the "work of a craftsman" as it is a delicate process that is affected by many variables, requiring the acquisition of specific and specialist know-how.

While islet isolation procedures are similar in both animal models and humans, the islets from different species have distinct anatomical and functional characteristics. Therefore, both common and unique features between species must be taken into account when isolating these structures in order to: (1) avoid inconsistencies introduced by the procedure used for islet isolation; (2) optimize the conditions of the isolation procedure and its outcome in terms of islet quality, as well as the time and cost of isolation; and (3) facilitate the translation of procedures developed in animal models to clinical settings.

This book, aimed at experts and beginners, addresses the challenges, pitfalls, and particularities of clinical islet isolation and those associated with their isolation from model animals. The book reviews the state of the art in this field, assessing the similarities and differences between human and animal islets, and how these influence their isolation, enabling strategies to be devised that can be translated to the clinic.

The first chapter is an introduction to the historical background of islet isolation, a fascinating story that has progressed hand in hand with that of islet transplantation. Indeed, our current mastery of both these processes can be expected to pave the way for the development of future cell therapies that will address the shortage of donor islets to treat diabetes.

In the following chapters, the procedures to isolate islets from mice, pigs, and nonhuman primates are reviewed, the main animal models used in preclinical studies and translational approaches. Working with mice has many advantages (they are relatively economic to maintain and easy to work with, they reproduce rapidly and in suitable numbers, etc.), and this species represents a true workhorse in this field of research. Porcine islets represent a very interesting model system, providing raw material for xenotransplantation, while nonhuman primates are the closest phylogenetic animal model to humans, the two species sharing a similar islet cytoarchitecture. As such, data obtained in nonhuman primates has a strong translational potential.

The lasts chapters focus on clinical islet isolation and all the processes and facilities required to establish a Clinical Islet Program: the donor organ, the effect of BMI, cold ischemia time, pancreas preservation, the procedure of islet isolation, islet culture, etc.

Finally, I would like to express my gratitude to all the authors who have contributed to this book and for the time and effort they dedicated to make it possible. I feel especially indebted to Dr. Juan Domínguez-Bendala for his assistance and his constant support. In addition, I would also like to thank Meran Owen and Tanja Koppejan at Springer for their invaluable assistance during the preparation of the book.

Leioa, Biscay, Spain Miriam Ramírez-Domínguez

Contents

Contributors

Midhat H. Abdulreda Diabetes Research Institute/Department of Surgery, University of Miami Leonard M. Miller School of Medicine, Miami, FL, USA

Per-Olof Berggren Diabetes Research Institute/Department of Surgery, University of Miami Leonard M. Miller School of Medicine, Miami, FL, USA

The Rolf Luft Research Center for Diabetes and Endocrinology, Karolinska Institutet Karolinska University Hospital L1, Stockholm, Sweden

Dora M. Berman Diabetes Research Institute, University of Miami Leonard M. Miller School of Medicine, Miami, FL, USA

Daniel Brandhorst Nuffield Department of Surgical Sciences, University of Oxford, Oxford, UK

Oxford Centre for Diabetes, Endocrinology and Metabolism, Oxford, UK

Heide Brandhorst Nuffield Department of Surgical Sciences, University of Oxford, Oxford, UK

Oxford Centre for Diabetes, Endocrinology and Metabolism, Oxford, UK

Over Cabrera Diabetes Research Institute/Department of Surgery, University of Miami Leonard M. Miller School of Medicine, Miami, FL, USA

Alejandro Caicedo Diabetes Research Institute/Department of Surgery, University of Miami Leonard M. Miller School of Medicine, Miami, FL, USA

Division of Endocrinology, Diabetes and Metabolism, Department of Medicine, University of Miami Leonard M. Miller School of Medicine, Miami, FL, USA

Yi Vee Chew National Pancreas and Islet Transplant Laboratories, The Westmead Institute for Medical Research, Westmead, NSW, Australia

Wayne J. Hawthorne National Pancreas and Islet Transplant Laboratories, The Westmead Institute for Medical Research, Westmead, NSW, Australia

Department of Surgery, Westmead Clinical School, Westmead Hospital, University of Sydney, Westmead, NSW, Australia

Paul R.V. Johnson Nuffield Department of Surgical Sciences, University of Oxford, Oxford, UK

Oxford Centre for Diabetes, Endocrinology and Metabolism, Oxford, UK

Oxford NIHR Biomedical Research Centre, Oxford, UK

Miriam Ramírez-Domínguez Laboratory of Cell Therapy of Diabetes, Department of Pediatrics, Faculty of Medicine and Odontology, Hospital Cruces, University of the Basque Country (UPV/EHU), Leioa, Biscay, Spain

Rayner Rodriguez-Diaz Division of Endocrinology, Diabetes and Metabolism, Department of Medicine, University of Miami Leonard M. Miller School of Medicine, Miami, FL, USA

The Rolf Luft Research Center for Diabetes and Endocrinology, Karolinska Institutet, Karolinska University Hospital L1, Stockholm, Sweden

Lindy Williams National Pancreas and Islet Transplant Laboratories, The Westmead Institute for Medical Research, Westmead, NSW, Australia

Historical Background of Pancreatic Islet Isolation

1

Miriam Ramírez-Domínguez

Abstract

Until the discovery of insulin in the twentieth century, diabetes mellitus was a mortal disease with an unclear origin and physiology. Despite the appearance of the concept in an Egyptian papyrus dated c.1550 BC, and the documentation of its study by ancient Chinese, the term "diabetes" was only coined by the Greek Aretaeus in the second century AD. In Europe, the study of diabetes was largely ignored until the seventeenth century, when the characteristic sweet flavor of diabetic urine was first described. However, the link between diabetes and the pancreas was not discovered until 1889 by Minkowski and von Mering, long after the first description of the pancreatic islets by Paul Langerhans in 1869. One of the most significant milestones in the field was the discovery of insulin by Banting and collaborators in 1922, which led to the therapeutic development of insulin administration as a life-saving intervention for type 1 diabetic patients. On the other hand, the isolation of islets was first reported by Bensley in 1911, a critical technical achievement that paved the way for clinical islet transplantation. Here we discuss the history of islet isolation, since the firsts studies of diabetes by ancient civilizations to the birth and parallel evolution of islet isolation and transplantation.

Keywords

Diabetes • History of diabetes • Insulin • Discovery of insulin • Islets of Langerhans • Islet isolation • Islet transplantation

1.1 The History of Diabetes Mellitus

Diabetes mellitus is defined by the American Diabetes Association as a "group of metabolic diseases characterized by hyperglycemia

M. Ramírez-Domínguez (✉)
Laboratory of Cell Therapy of Diabetes, Department of Pediatrics, Faculty of Medicine and Odontology, Hospital Cruces, University of the Basque Country (UPV/EHU), Barrio Sarriena, s/n 48940, Leioa, Biscay, Spain
e-mail: miriamrd@gmail.com

M. Ramírez-Domínguez (ed.), *Pancreatic Islet Isolation*, Advances in Experimental Medicine and Biology 938, DOI 10.1007/978-3-319-39824-2_1

resulting from defects in insulin secretion, insulin action or both" [1]. However, polyuric diseases were known over 3500 years ago (Table 1.1). The first mention to them appears in an Egyptian papyrus dating from c. 1550 BC, discovered by Georg Evers. The term "diabetes" was not given to what we now call type I diabetes until the second century AD, by the Greek Aretaeus. The origin of "diabetes" is in the Greek word for a siphon, since Aretaeus said that "the fluid does not remain in the body, but uses the man's body as a channel whereby to leave it".

Table 1.1 Milestones in the history of diabetes

Ebers papyrus (Egipt, 1500 BC)	Polyuric diseases
Charak and Sushrut (India, 5th century BC)	Sweet urine diseases
"The Yellow Emperor's Canon on the Traditional Chinese Medicine" (China, 4th century BC)	Sweet urine disease
Aretaeus (Cappadocia, 2nd century AD)	Polyuric state called "diabetes"
Avicenna (Arabia, 10th century, AD)	Sweet urine disease
Thomas Willis (England, 1674)	Sweet urine disease
Matthew Dobson (England, 1776)	Hyperglycemia in urine and serum
John Rollo (England, 1797)	Diabetes is called "diabetes mellitus". Design of the "animal diet"
Michel Chevreul (France, 1815)	The sugar in diabetic urine is glucose
Claude Bernard (France, 1846–1848)	Glucose is stored in the liver as "glycogen" and released into the blood during fasting
Paul Langerhans (Germany, 1869)	Description of pancreatic islets
Etienne Lanceraux (France, 1880)	Classificaton of diabetes ("diabète maigre" and "diabète gras")
Oskar Minkowski and Josef von Mering (Germany, 1890)	Link between diabetes and the pancreas. Pancreatectomy causes diabetes in dogs
Gustave Edouard Laguesse (France, 1893)	The "internal secretions" of the pancreas are produced by the "islets of Langerhans"
Jean de Meyer (Belgium, 1909)	"Internal secretions" of the pancreas are called "insuline"
Frederick Banting, Charles Best, JJR Macleod and James Collip (Canada, 1922)	Discovery of insulin

Between 400 and 500 BC, the Hindu physicians Charak and Sushrut were probably the first to identify the sweetness of diabetic urine. In parallel, around 400 BC, sweet urine disease was mentioned in the oldest Chinese medical book, "The Yellow Emperor's Canon on the Traditional Chinese Medicine". This was also recognized by Arab physicians in medical texts from the ninth to eleventh centuries AD, especially in the medical encyclopedia written by Avicenna.

In Europe, the disease was largely ignored until Thomas Willis wrote "Diabetes, or the Pissing Evil" in 1674 [2]. He stated that the urine was "wonderfully sweet like sugar or honey" but he did not consider that the cause might be the content of sugar in it.

In 1776 Matthew Dobson described hyperglycemia for the first time. He observed the sweet flavor of urine and serum of one of his patients and he concluded that the kidneys excreted sugar that previously existed in the serum of the blood [3].

Some years later, John Rollo, a surgeon trained in Edinburgh, was the first to add the adjective "mellitus" to diabetes, from the Latin word meaning "honey". He also famously developed a diet (the "animal diet") [4] to treat diabetic patients, which became the standard treatment in the nineteenth century. It was a diet based on animal food, since it was thought that sugar was formed from vegetables in the stomach.

In 1815, the French chemist Michel Chevreul proved that the sugar in diabetic urine was glucose [5]. Later, in the middle of the nineteenth century, the method to diagnose diabetes evolved from tasting urine to chemical tests for reducing agents such as glucose. At the beginning, the measurement of glycemia required so much blood that it was rarely practiced in either clinical care or research. But in 1913, the Norwegian physician Ivar Christian Bang introduced a micromethod which led to the development of the glucose tolerance tests.

Until the first half of the nineteenth century, it was thought that sugar could only be found in plants, and therefore, the sugar could be found in animals when they broke down food of plant origin. But Claude Bernad discovered between 1846 and 1848 that glucose was also present in the blood of animals, even when they starved. He also discovered that there was a substance similar to starch in the liver that converted to sugar, and he called this "glycogen" (sugar-forming) [6]. His theory was that sugar was absorbed by the intestine and then it was converted into glycogen in the liver, to be constantly released into the blood during fasting.

In 1869, Paul Langerhans discovered with his doctoral thesis the existence of clusters of cells in the pancreas, despite their function was unknown [7]. However, the link between diabetes and the pancreas was not discovered until 1889 by Minkowski and von Mering. While studying fat metabolism, they serendipitously realized that the cause of constant urination in a dog was the pancreatectomy they had performed. Upon testing the dog's urine, they hypothesized that the pancreas produced an internal secretion that regulated carbohydrate metabolism [8]. Then, in 1893, Gustave Laguesse hypothesized that the "internal secretions" of the pancreas were produced by the "islets of Langerhans" [9]. In 1909, the Belgian Jean de Meyer coined the term "insuline" to refer to the "internal secretions" of the pancreas, from the Latin word for "island" [10].

However, the link between pancreas and diabetes was not immediately adopted. For 20 years, the scientific community debated about the subtypes of diabetes and its pathogenesis. In fact, in 1880, Etienne Lancereaux distinguished between "*diabète maigre*" and "*diabète gras*" [11] in patients lean and obese, establishing the earliest classifications of the disease.

1.1.1 Discovery of Insulin

There were many attempts to isolate the "internal secretions" of the pancreas during the first two decades of the twentieth century. The ones who came closer were Georg Zuelzer in 1907 [12]; Ernest Scott in 1911 [13]; John Murlin in 1913 [14]; Israel Kleiner in 1919 [15], and Nicholas Paulesco in 1920–1921 [16]. However, their efforts were unsuccessful due to the inactivation of the extracts or problems with impurities.

It was not until October 1920 that Frederick Banting, a young orthopedic surgeon, got inspired while reading an article to prepare a lecture about the pancreatic islets of Langerhans and diabetes. He hypothesized that ligation of the pancreatic ducts before the extraction of the organ would destroy the acinar tissue, the enzyme-secreting compartment of the pancreas, while the islets of Langerhans would remain intact and able to produce the internal secretion regulating sugar metabolism. He thought that the previous attempts in extracting the "internal secretions" failed due to the destructive action of trypsin released by the pancreas.

His hypothesis was based on previous knowledge developed by Ssobolew in 1902 [17] and Opie in 1900 [18]. Ssobolew had shown that ligation of the pancreatic ducts was linked to a gradual atrophy and destruction of the acini, while the islets remained intact. Opie, on the other hand, showed islet degeneration associated with diabetes, implying that islets were responsible for an internal secretion of the pancreas that was essential for the metabolism of carbohydrates.

Banting subsequently approached J.J.R. Macleod, at the University of Toronto, who was a leading authority on carbohydrate metabolism, and asked for laboratory space to develop his hypothesis. Macleod accepted and Banting started working there with an assistant student, Charles Best. They followed Macleod's instructions to prepare extracts of atrophied pancreas from dogs pancreatectomized to become diabetic and then they injected them the extract. Some months later, Banting realized they could also obtain active extracts more easily and capable of large-scale production using beef pancreata from the abattoir. He recalled that Laguesse found that islet cells were more abundant than acini in fetal and newborn animals than in adult animals, and therefore their extracts would be free from trypsin activity.

Later, they optimized the extraction procedure with the participation of James B. (Bert) Collip, a biochemist who was in a sabbatical leave visiting the University of Toronto. On January 11th 1922, the first clinical trial took place, administering the extract to a 14-year-old diabetic patient, Leonard Thompson, with no clinical benefit observed. However, on January 23rd and for the next 10 days, another extract was administered again to the same patient, with clinical improvement and complete elimination of glycosuria and ketonuria.

At first they named the extract "isletin", but Macleod suggested to call it "insulin", unaware that de Meyer had previously suggested "insuline". They started the large scale production in collaboration with Eli Lilly, and in 1923 Banting and Macleod received jointly the Nobel Prize for Physiology or Medicine, sharing it later with Best and Collip [19].

1.2 The History of Islet Isolation

With the discovery of insulin, diabetes became a chronic illness with severe complications instead of being a mortal disease. On one hand, with the discovery of insulin, the interest in replacement strategies with pancreatic fragments decreased. On the other, the improvements in islet isolation in animal models had an important impact in islet isolation and transplantation in humans, and since then, these two fields have evolved in parallel.

Before the discovery of insulin there were researchers who worked in the hypothesis that transplanting pancreatic fragments into diabetic animals could cure the disease, since they thought that there was a substance, maybe located in the pancreas, that destroyed the sugar.

The first ones reporting a successful trial were Oscar Minkowski and Joseph von Mering. In 1892, they transplanted autologous pancreatic fragments subcutaneously in a pancreatectomized diabetic dog, demonstrating transient improvement of glycosuria [20].

The next year, P. Watson Williams and surgeon William H. Harsant performed in the UK the first subcutaneous xenotransplantation of three fresh sheep pancreatic fragments in a 15-year old boy, who eventually died [21]. For the next few years, the scientific community focused on demonstrating that the "internal secretions" of the pancreas could be beneficial for the evolution of the disease if transplanted in alternative sites to the subcutaneous space [22–27].

In 1916, the British surgeon Frederick Charles Pybus, noticing that previous attempts with xenogeneic material had failed, performed an allogeneic transplant [28]. He transplanted a human pancreas immediately after the death of the donor, placing it in the abdominal space of two diabetic patients. In one of them he achieved a transient reduction in glycosuria, but there was no reversal of diabetes and both of them died. The principles of immune rejection in transplantation were still unknown.

In 1902, the Russian doctor Leonid W. Ssobolew suggested the idea of physically separating the exocrine tissue from the endocrine tissue before the transplant [17] according to the hypothesis that the former could impair the viability and function of the latter. This idea was first brought to fruition in 1911 with the pioneering work of R.R. Bensley, with the staining of islets with neutral red and the hand-picking method [29] (Table 1.2).

In 1964, Dr. Hellerström started the development of islet isolation techniques by microscope microdissection of islets from the pancreas of obese hyperglycemic mice, with poor results in yield and quality [30]. However, in 1965, Dr. Moskalewski introduced for the first time the use of collagenase in islet isolation [31]. He isolated minced guinea pig pancreas with bacterial collagenase from *Clostridium histolyticum* to release islet clusters from the exocrine tissue, despite widespread islet destruction due to the activity of the enzyme.

This method was improved by Drs. Paul E. Lacy and Mery Kostianovsky at Washington University in Saint Louis [32], taking advantage of the pancreatic anatomy and introducing intra-ductal injection of cold saline buffer to distend the pancreas and increase the pancreas surface to the action of collagenase to enhance islet release. They also performed an enzymatic digestion after harvesting and mincing the pancreas, with final islet hand-picking under the dissecting

Table 1.2 Milestones in the history of islet isolation

R. R. Bensley (USA, 1911)	Islet staining with neutral red and hand-picking
C. Hellerström (Sweden, 1964)	Microscope microdissection of islets
S. Moskalewski (Poland,1965)	Use of collagenase in mouse islet isolation
P. E. Lacy and M. Kostianovsky (USA, 1967)	Pancreas distention by intra-ductal injection of cold saline buffer
A. Lindall (USA, 1969)	Islet purification by Ficoll density gradient
A. Horaguchi and R. Merrell (USA, 1981)	Design of a new system to perfuse the pancreas
M. Gotoh (Japan, 1985)	Pancreas distention by intra-ductal injection of collagenase
Camillo Ricordi (USA, 1988)	Design of the "Ricordi chamber"
S. Lake (UK, 1989)	Introduction of the COBE 2991 in human islet isolation
Marketing of Liberase HI by Roche (USA, 1994)	Optimization of human islet enzymatic dissociation
J. Lakey (Canada, 1999)	Introduction of a recirculating controlled perfusion system in human islet isolation

microscope. However, it was not until 1985 that the isolation method in rodents was perfected by Gotoh et al. who performed intra-ductal injection of collagenase, instead of buffer [33].

However, hand-picking isolation was a tedious procedure, which was not feasible for large-scale islet isolation due to poor yield. Alternative purification procedures, such as density gradient purification, were thus developed. The first density gradients were based on sugar or albumin. Ficoll was later introduced by Arnold Lindall et al. at the University of Minnesota [34]. Ficoll is a high molecular weight polymer of sucrose, which improved islet purification from acinar tissue. However, although high yields were obtained with Ficoll, the cells were not functional, since Ficoll was prepared with a high concentration of sucrose and was hyperosmolar, impairing insulin secretion. Dr. Lacy further improved this method dialyzing and lyophilizing Ficoll, with positive results. He established a standardized methodology in rodent islet isolation and made routine rodent islet transplantation studies feasible [35]. He established two different phases in the procedure: islet dissociation and islet purification.

In 1972, Ballinger and Lacy observed an improvement (but no complete reversal) of experimental diabetes in rats, transplanting 400–600 islets intraperitoneally or intramuscularly [36]. Just one year later, Reckard and Barker achieved the reversal of experimental diabetes for the first time, transplanting a larger number of islets (800–1200) intraperitoneally [37].

In 1973, Charles Kemp performed the first study linking transplantation site and outcome in rats. With only 400–600 transplanted islets, there was a complete reversal of diabetes in 24 h when delivering them in the liver, but no success was achieved when transplanting the same number of islets into the peritoneal cavity or subcutaneously [38]. From that moment on, the liver was accepted as the gold standard place for transplantation in rodent models as well as in the clinical setting. The advantages of the liver as an ectopic transplantation site are its high vascularity and its proximity to islet nutrients and growth factors. Physiologically, it is also a place of delivery of insulin [39]. However, it has been reported that a 60 % of islets transplanted in the liver die shortly after transplantation [40]. The main reason is that the hepatic oxygen tension is low, even lower than pancreas, and islets recently implanted lack proper vasculature and die due to chronic hypoxia. Besides, it is an organ with a high metabolic activity, producing massively radicals and metabolites that generate an adverse cytokine/chemokine environment for islets, and there is local inflammatory activity, which affects long-term graft survival. Therefore, the islet community is currently focusing efforts in finding an alternative optimal transplantation ectopic site for islets [41].

Once demonstrated that diabetes could be reversed by transplantation in rodents, the next step was to translate this knowledge to human islets isolation and transplantation. However, there are intrinsic differences between rodent and human islets [42–45], which makes it difficult to extrapolate the techniques. Therefore, islet researchers started preclinical assays with dogs, considering that the canine pancreas is more similar to the human one in density and fibrosity.

The translation of techniques from rodents to large animals (dog, nonhuman primate and human) was not easy, and the cell preparations were not completely pure until 1977 [46–48]. In 1976, Mirkovitch and Campiche were the first to reverse diabetes in pancreatectomized dogs with partially digested pancreatic tissue autotransplanted in the spleen [49]. In humans, Najarian et al. [50, 51] also transplanted partially purified pancreatic fragments. However, the metabolic control was poor, the immunosuppresion inadequate, the endocrine mass transplanted was not enough and there were complications derived from the insufficient degree of purification achieved. Actually, it has been reported that intrasplenic transplantation of impure or partially purified tissue may cause morbidity, splenic rupture and portal vein thrombosis, despite achieving insulin independence [52].

During that period, islet isolation procedures were further improved by some researchers. Horaguchi and Merrell, at Standford University, designed a system to perfuse the pancreas with collagenase once the pancreatic duct was cannulated. This was followed by a step of mechanical dissociation and digestion with collagenase, first, and trypsin, second, with a third step of filtration through a 400 μm mesh, yielding a 57 % of islet recovery [53].

Dr. Mintz et al. [54] and Dr. Gray et al. [55] further developed a new method for islet isolation improving the dissociation of the pancreas by passing the digest through different graded needles to separate the islets of the exocrine tissue, and next purifying by filtration and application of density gradients. With these modifications, the purity obtained with human pancreas reached the 20–40 % [55, 56]. Although there was still some islet destruction due to the enzymatic activity, this method allowed for the successful islet isolation from pigs [57] non-human primates [58] and humans [55].

A milestone in the field of islet isolation and transplantation was the invention of the Ricordi chamber, in 1988 [59]. Camillo Ricordi had joined Dr. Lacy's team 2 years before and he introduced a method to improve the digestion and dissociation of human pancreas that was less traumatic than previous methods. He designed a dissociation/filtration chamber, called the Ricordi chamber, which consisted in an upper conical part separated by a 500 μm mesh from the lower cylindrical part with stainless steel spheres. The pancreatic tissue was placed in the lower part, and it was digested by a combination of enzymatic digestion at 37 °C and gentle mechanical agitation of the chamber. There was a continuous flow between the heating system and the chamber, through a peristaltic pump. When islets were released, they were filtered through the mesh and collected from the upper part of the chamber. The point when the collection started was decided after sequential sampling, therefore avoiding overdigestion. This method was a success and since then, the Ricordi chamber has been the gold standard for human and large animal pancreatic islet isolation all over the world.

In that same year, Dr. Lake et al. reported a method that allowed large-scale purification of human islets, with the COBE 2991 processor [60]. This device was originally used to process bone marrow but allowed the purification of a single human pancreas by Ficoll in a sterile system. This is still the method used currently to process large animal pancreata.

In 1994, an enzyme blend that revolutionized human islet isolation and clinical islet transplantation was marketed. It was Liberase HI (Roche, Indianapolis, USA), a low-endotoxin enzyme which was the first enzyme designed especially for human islet isolation. It showed superior enzymatic action in comparison with the traditional enzyme preparation (collagenase P) [61]. However, Liberase was removed from the market in 2007 due to the potential risk of transmitting bovine spongiform encephalopathy to patients because this enzyme is isolated from *Clostridium histolyticum* grown in media containing brain-heart infusion broth [62, 63]. As this was the enzyme of choice in the field (used in 77 % of cases, based on CITR data [64]), the withdrawal of Liberase HI from the market resulted in a reduction in the number of clinical islet transplantations [65]. Nowadays there are recombinant alternatives that circumvent the above risks.

In 1999, Dr. Lakey et al. reported a recirculating controlled perfusion system that allowed for the control of the digestion temperature [66]. This resulted in a more effective delivery of the enzyme, yielding more islets and facilitating human islet recovery and survival in comparison with syringe loading.

1.3 Concluding Remarks

Our understanding of diabetes has evolved tremendously from the first documentation of the disease by ancient Egyptians until the discovery of insulin in the twentieth century and the development of current cell replacement therapies. Islet transplantation is a long and storied field of research that has gone hand in hand with progress in islet isolation. Our current mastery of both processes is expected to pave the way for the next generation of cell therapies for diabetes, which will address the shortage of cadaveric islets by employing stem cell-derived insulin-producing cells.

References

1. American Diabetes Association. Diagnosis and classification of diabetes mellitus. Diabetes Care. 2011;34(Suppl 1):S62–9.
2. Willis T. Pharmaceutic Rationalis; sive, Diatriba de Medicamentorum Operationibus in Humano Corpore. Oxford: John Anthony Huguetan; 1674.
3. Dobson M. Experiments and observations on the urine in diabetes. Med Obs Inq. 1776;5:289–316.
4. Rollo J. An account of two cases of the diabetes mellitus. London: Printed by T. Gillet, for C. Dilly; 1797.
5. Chevreul M. Note sur le sucre du diabète. Ann Chim. 1815;95:319–20.
6. Olmsted J. Claude Bernard, 1813–1878: a pioneer in the study of carbohydrate metabolism. Diabetes. 1953;2(2):162–4.
7. Langerhans P. Beiträge zur mikroskopichen Anatomie der Bauchspeicheldrüse (dissertation). Berlin: Gustave Lange; 1869.
8. Mering J, Minkowski O. Diabetes mellitus nach Pankreasexstirpation. Archiv Exp Pathol Pharmakol. 1890;26(5–6):371–87.
9. Laguesse G. Sur la formation des ilots de Langerhans dans le pancreas. C R Seances Mem Soc Biol. 1893;45:819–20.
10. De Meyer J. Sur la signification physiologique de la secretion interne du pancreas. Zentralbl Physiol. 1904;18:S826.
11. Lancereux E. Le diabète maigre: ses symptomes son evolutions, son prognostic et son traitement. Union Méd Paris. 1880;20:205–11.
12. Zuelzer G. Experimentelle Unterschungen Über den Diabetes. Berlin Klin Wochenschr. 1907;44:474–5.
13. Richards D. The effect of pancreas extract on depancreatized dogs: Ernest L. Scott's Thesis of 1911. Perspect Biol Med. 1966;10(1):84–95.
14. Murlin J, Kramer B. The influence of pancreatic and duodenal extracts on the glycosuria and the respiratory metabolism of depancreatized dogs. Exp Biol Med. 1913;10(5):171–3.
15. Kleiner I. The action of intravenous injection of pancreas emulsions in experimental diabetes. J Biol Chem. 1919;40:153–70.
16. Paulesco N. Recherche sur le role du pancreas dans l'assimilation nutritive. Arch Int Physiol. 1921;17:85–109.
17. Ssobolew L. Zur normalen und pathologischen Morphologie der inneren Secretion der Bauchspeicheldrüse. Arch f Pathol Anat. 1902;168(1):91–128.
18. Opie E. The relation of diabetes mellitus to lesions of the pancreas. Hyaline degeneration of the islands of Langerhans. J Exp Med. 1901;5(5):527–40.
19. Rosenfeld L. Insulin: discovery and controversy. Clin Chem. 2002;48(12):2270–88.
20. Minkowski O. Weitere mittheilungen Uber den Diabetes mellitus nach exstirpation des pancreas. Berl Klin Wochenschr. 1892;29:90–4.
21. Watson-Williams P. Notes on diabetes treated with extract and by grafts of sheeps pancreas. Br Med J. 1894;2:1303–4.
22. Piemonti L, Pileggi A. 25 years of the Ricordi automated method for islet isolation. CellR4. 2013;1(1):e128.
23. Alessandri R. Innesti di tessuti viventi adulti ed embrionali in alcuni organi del corpo. Policlinico. 1896;III-C:253–71.
24. Kyrle J. Über die Degenerationsvorgange im tierischen Pankreas. Arch Mikrosk Anat. 1908;72(1):141–60.
25. Ottonlenghi D. Sur la transplantation du pancreas. Arch Ital Biol. 1901;36:447.
26. Pratt J, Murphy F. Pancreatic transplantation in the spleen. J Exp Med. 1913;17:252.
27. Allan J. Transplantation of pancreas in diabetes. BMJ. 1903;1(2204):764.
28. Pybus F. Notes on suprarenal and pancreatic grafting. Lancet. 1924;204(5272):550–1.

29. Bensley R. Studies on the pancreas of the guinea pig. Am J Anat. 1911;12(3):297–388.
30. Hellerstrom C. A method for the microdissection of intact pancreatic islets of mammals. Eur J Endocrinol. 1964;45(1):122–32.
31. Moskalewski S. Isolation and culture of the islets of Langerhans of the guinea pig. Gen Comp Endocrinol. 1965;5(3):342–53.
32. Lacy P, Kostianovsky M. Method for the isolation of intact islets of Langerhans from the rat pancreas. Diabetes. 1967;16(1):35–9.
33. Gotoh M, Maki T, Kiyoizumi T, Satomi S, Monaco A. An improved method for isolation of mouse pancreatic islets. Transplantation. 1985;40(4):437.
34. Lindall A, Steffes M, Sorenson R. Immunoassayable insulin content of subcellular fractions of rat islets 1. Endocrinology. 1969;85(2):218–23.
35. Younoszai R, Sorensen R, Lindall A. Homotransplantation of isolated pancreatic islets. Diabetes. 1970;19(Supp.1):406.
36. Ballinger W, Lacy P. Transplantation of intact pancreatic islets in rats. Surgery. 1972;72(2):175–86.
37. Reckard C, Ziegler M, Barker C. Physiological and immunological consequences of transplanting isolated pancreatic islets. Surgery. 1973;74(1):91.
38. Kemp C, Knight M, Scharp D, Ballinger W, Lacy P. Effect of transplantation site on the results of pancreatic islet isografts in diabetic rats. Diabetologia. 1973;9(6):486–91.
39. Lakey JRT, Burridge PW, Shapiro AMJ. Technical aspects of islet preparation and transplantation. Transpl Int. 2003;16:631–2.
40. Bennet W, Sundberg B, Groth CG, Brendel MD, Brandhorst D, Brandhorst H, et al. Incompatibility between human blood and isolated islets of Langerhans: a finding with implications for clinical intraportal islet transplantation? Diabetes. 1999;48(10):1907–14.
41. Merani S, Toso C, Emamaulee J, Shapiro AMJ. Optimal implantation site for pancreatic islet transplantation. Br J Surg. 2008;95:1449–61.
42. Brissova M, Fowler MJ, Nicholson WE, Chu A, Hirshberg B, Harlan DM, Powers AC. Assessment of human pancreatic islet architecture and composition by laser scanning confocal microscopy. J Cyotchem Histochem. 2005;53(9):1087–97.
43. Cabrera O, Berman DM, Kenyon NS, Ricordi C, Berggren PO, Caicedo A. The unique cytoarchitecture of human pancreatic islets has implications for islet cell function. PNAS. 2006;103(7):2334–9.
44. Bonner-Weir S, Sullivan BA, Weir GC. Human islet morphology revisited: human and rodent islets are not so different after all. J Histochem Cytochem. 2015;63(8):604–12.
45. Dolensek J, Rupnik MS, Stozer A. Structural similarities and differences between the human and the mouse pancreas. Islets. 2015;7(1):e1024405.
46. Lorenz D, Lippert H, Panzig E, Köhler H, Koch G, Tietz W, et al. Transplantation of isolated islets of Langerhans in diabetic dogs. III. Donor selection by mixed lymphocyte reaction and immunosuppressive treatment. J Surg Res. 1979;27(3):205–13.
47. Kolb E, Ruckert R, Largiader F. Intraportal and intrasplenic autotransplantation of pancreatic islets in the dog. Eur Surg Res. 1977;9(6):419–26.
48. Kretschmer G, Sutherland D, Matas A, Steffes M, Najarian J. The dispersed pancreas: transplantation without islet purification in totally pancreatectomized dogs. Diabetologia. 1977;13(5):495–502.
49. Mirkovitch V, Campiche M. Successful intrasplenic autotransplantation of pancreatic tissue in totally pancreatectomised dogs. Transplantation. 1976;21(3):265–9.
50. Najarian J, Sutherland D, Matas A, Steffes M, Simmons R, Goetz F. Human islet transplantation: a preliminary report. Transplant Proc. 1977;9(1):233–6.
51. Sutherland D, Matas A, Goetz F, Najarian J. Transplantation of dispersed pancreatic islet tissue in humans: autografs and allografts. Diabetes. 1980;29(Supplement 1): 31–44.
52. Gores P, Najarian J, Stephanian E, Lloveras J, Kelley S, Sutherland D. Transplantation of unpurified islets from single donors with 15-deoxyspergualin. Transplant Proc. 1994;26(2):574.
53. Horaguchi A, Merrell R. Preparation of viable islet cells from dogs by a new method. Diabetes. 1981;30(5):455–8.
54. Noel J, Rabinovitch A, Olson L, Kyriakides G, Miller J, Mintz D. A method for large-scale, high-yield isolation of canine pancreatic islets of Langerhans. Metabolism. 1982;31(2):184–7.
55. Gray D, McShane P, Grant A, Morris P. A method for isolation of islets of Langerhans from the human pancreas. Diabetes. 1984;33(11):1055–61.
56. Alejandro R, Mintz D, Noel J, Latif Z, Koh N, Russell E. Islet cell transplantation in type I diabetes mellitus. Transplant Proc. 1987;19(1 Pt.3):2359–61.
57. Ricordi C, Finke E, Lacy P. A method for the mass isolation of islets from the adult pig pancreas. Diabetes. 1986;35(6):649–53.
58. Gray D, Warnock G, Sutton R, Peters M, McShane P, Morris P. Successful autotransplantation of isolated islets of Langerhans in the cynomolgus monkey. Br J Surg. 1986;73(10):850–3.
59. Ricordi C, Lacy P, Finke E, Olack B, Scharp D. Automated method for isolation of human pancreatic islets. Diabetes. 1988;37(4):413–20.
60. Lake S, Bassett P, Larkins A, Revell J, Walczak K, Chamberlain J, et al. Large-scale purification of human islets utilizing discontinuous albumin gradient on IBM 2991 cell separator. Diabetes. 1989;38(Supplement_1):143–5.
61. Rheinheimer J, Ziegelmann P, Carlessi R, Reck L, Bauer A, Leitão C, et al. Different digestion enzymes used for human pancreatic islet isolation: a mixed treatment comparison (MTC) meta-analysis. Islets. 2014;6(4):e977118.

62. Jahansouz C, Jahansouz C, Kumer S, Brayman K. Evolution of β-cell replacement therapy in diabetes mellitus: islet cell transplantation. J Transplant. 2011;2011:1–21.
63. Alejandro R, Barton F, Hering B, Wease S. 2008 update from the collaborative islet transplant registry. Transplantation. 2008;86(12):1783–8.
64. The CITR Coordinating Center and Investigators. The collaborative islet transplant registry 2009 annual report. 2009, http://www.citregistry.org/reports/reports.htm
65. Barton F, Rickels M, Alejandro R. Improvement in outcomes of clinical islet transplantation: 1999–2010. Diabetes Care. 2012;35(7):1436.
66. Lakey J, Warnock G, Ao Z, Shapiro A, Korbutt G, Kneteman N, et al. Intraductal collagenase delivery into the human pancreas using syringe loading or controlled perfusion. Transplant Proc. 1998;30(2):359.

The Different Faces of the Pancreatic Islet

2

Midhat H. Abdulreda, Rayner Rodriguez-Diaz, Over Cabrera, Alejandro Caicedo, and Per-Olof Berggren

Abstract

Type 1 diabetes (T1D) patients who receive pancreatic islet transplant experience significant improvement in their quality-of-life. This comes primarily through improved control of blood sugar levels, restored awareness of hypoglycemia, and prevention of serious and potentially life-threatening diabetes-associated complications, such as kidney failure, heart and vascular disease, stroke, nerve damage, and blindness. Therefore, beta cell replacement through transplantation of isolated islets is an important option in the treatment of T1D. However, lasting success of this promising therapy depends on durable survival and efficacy of the transplanted islets, which are directly influenced by the islet isolation procedures. Thus, isolating pancreatic islets with consistent and reliable quality is critical in the clinical application of islet transplantation.

Quality of isolated islets is important in pre-clinical studies as well, as efforts to advance and improve clinical outcomes of islet transplant ther-

M.H. Abdulreda (✉)
Diabetes Research Institute/Department of Surgery, University of Miami Leonard M. Miller School of Medicine, Miami, FL, USA
e-mail: mabdulreda@miami.edu

R. Rodriguez-Diaz
Division of Endocrinology, Diabetes and Metabolism, Department of Medicine, University of Miami Leonard M. Miller School of Medicine, Miami, FL, USA

The Rolf Luft Research Center for Diabetes and Endocrinology, Karolinska Institutet, Karolinska University Hospital L1, Stockholm SE-171 76, Sweden

O. Cabrera
Diabetes Research Institute/Department of Surgery, University of Miami Leonard M. Miller School of Medicine, Miami, FL, USA

A. Caicedo
Diabetes Research Institute/Department of Surgery, University of Miami Leonard M. Miller School of Medicine, Miami, FL, USA

Division of Endocrinology, Diabetes and Metabolism, Department of Medicine, University of Miami Leonard M. Miller School of Medicine, Miami, FL, USA

P.-O. Berggren
Diabetes Research Institute/Department of Surgery, University of Miami Leonard M. Miller School of Medicine, Miami, FL, USA

The Rolf Luft Research Center for Diabetes and Endocrinology, Karolinska Institutet, Karolinska University Hospital L1, Stockholm SE-171 76, Sweden

M. Ramírez-Domínguez (ed.), *Pancreatic Islet Isolation*, Advances in Experimental Medicine and Biology 938, DOI 10.1007/978-3-319-39824-2_2

apy have relied heavily on animal models ranging from rodents, to pigs, to nonhuman primates. As a result, pancreatic islets have been isolated from these and other species and used in a variety of in vitro or in vivo applications for this and other research purposes. Protocols for islet isolation have been somewhat similar across species, especially, in mammals. However, given the increasing evidence about the distinct structural and functional features of human and mouse islets, using similar methods of islet isolation may contribute to inconsistencies in the islet quality, immunogenicity, and experimental outcomes. This may also contribute to the discrepancies commonly observed between pre-clinical findings and clinical outcomes. Therefore, it is prudent to consider the particular features of pancreatic islets from different species when optimizing islet isolation protocols.

In this chapter, we explore the structural and functional features of pancreatic islets from mice, pigs, nonhuman primates, and humans because of their prevalent use in nonclinical, preclinical, and clinical applications.

Keywords
Islet isolation • Islet transplantation • Type 1 diabetes • T1D • Type 2 diabetes • T2D • Islet cytoarchitecture • Islet vasculature • Islet microcirculation • Islet innervation • Sympathetic • Parasympathetic • Autocrine signaling • Paracrine signaling • Basement membrane • Neurotransmitter • Glutamate • GABA • ATP • Insulin • Glucagon • Somatostatin • Signaling hierarchy • Endocrine cells • Endocrine pancreas

2.1 Introduction

Diabetes is reaching pandemic proportions worldwide and is among the leading causes of morbidity and mortality. This is primarily due to serious complications associated with this devastating disease. Such complications include blindness, amputations, kidney failure, heart and vascular disease, stroke, nerve damage, and even birth defects [1–3].

Although the specific etiologies of either form of diabetes are still unknown [4, 5], it is well established that T1D results from the autoimmune destruction of the insulin-producing beta cells in the endocrine pancreas (i.e., the islets of Langerhans). T2D is thought to manifest in individuals with risk factors which include but not limited to genetic predisposition, obesity, and sedentary lifestyle [1, 6–15]. While lifestyle changes and therapeutic intervention may be effective in preventing and/or treating T2D [7, 9], treatment options in T1D are limited to insulin supplementation either in the form of injectable insulin or biological replacement of the insulin-producing beta cells [5].

Several options of beta cell replacement have been pursued in the last few decades. Regenerative approaches such as inducing proliferation of existing mature beta cells, differentiation of stem cells and/or trans-differentiation of other endocrine or non-endocrine cells into insulin-producing cells hold great promise in treating T1D [16–19]. But these approaches are yet to materialize into safe and reliable clinical applications. Transplantation offers another option of biological replacement but also has limitations. Limited availability of donor tissue remains a significant obstacle in transplantation therapies in general including that of pancreatic islets. Other limitations are associated with the required use of immunosuppressive drugs to prevent immune-mediated rejection; chronic systemic immunosuppression exposes transplant recipients to serious and potentially deadly side-effects and complications such as increased susceptibility to infections/sepsis and cancer development.

Although, immunosuppressive agents are continuously being improved and new ones are being developed to better protect the grafts while reducing their undesired side-effects, the health risks associated with chronic systemic immunosuppression remain high. Nonetheless, the risk to benefit consideration in many patients favor transplantation, especially where improvement in quality-of-life is expected. This has been well documented in transplant therapy in T1D diabetes patients [20–24].

T1D patients currently receive transplants either in the form of whole pancreas or isolated pancreatic islets. On one hand, whole pancreas transplantation achieves complete insulin independence in T1D patients, but it is highly invasive and is associated with high risk of complications including mortality. On the other hand, transplantation of isolated pancreatic islets is minimally invasive and has significantly less complications compared to whole pancreas transplant, but survival of the islet graft might be limited due to complications associated with the current clinical transplant site, the portal system of the liver. Nevertheless, hundreds of T1D patients have received islet transplants in the liver in the last two and a half decades in clinical trials [25]. These studies have shown that islet transplant recipients benefit from improved glycemic control and prevention of severe hypoglycemia as well as other diabetes-associated complications (see above). This improves the patients' quality of life significantly. Therefore, transplantation of isolated pancreatic islets has emerged as a promising therapy for T1D [23–25], and is on the verge of becoming standard-of-care in the United States and other countries.

As mentioned above, organ/tissue transplantation from non-related (i.e., allogeneic) donors is associated with risk of immune-mediated rejection of the allograft. As with other allotransplantations, recipients of islet allografts require life-long immunosuppression therapy to prevent rejection. It is also well established that the immunogenicity of transplanted pancreatic islets can play a key role in inflammation and anti-graft immunity in the immediate post-transplant period [26–28]. The immunogenicity of isolated islets is affected significantly by the isolation procedure [29–33]. Importantly, less immunogenic islets stand a better chance of survival and successful engraftment after transplantation [34]. This directly impacts on the success of islet transplantation and the clinical outcome. Therefore, efforts to optimize conditions for isolating pancreatic islets are constantly pursued in pre-clinical and clinical applications.

Pancreatic islets have been isolated from different species for a variety of purposes ranging from pre-clinical in vitro or in vivo studies in animals to transplantation into human patients. While islet isolation protocols and procedures may vary as described elsewhere in this book, they have been somewhat similar for isolating islets from mammals [35–37]. However, using similar methods to isolate islets from different species may contribute to severe inconsistency in islet yield and quality. Therefore, we dedicate this chapter to highlight different structural and functional features of islets from different species, which should be considered during optimization of islet isolation procedures. We focus primarily on pancreatic islets from mice, pigs, nonhuman primates, and humans because of their prevalent use in nonclinical, preclinical, and clinical applications. Mouse islets have been and are likely to remain the workhorses of islet biology research. Porcine islets are critical in the field of xenotransplantation as they provide a potentially limitless source of pancreatic islets for transplantation into T1D patients. Nonhuman primates (NHP) are a reliable surrogate for human islets in preclinical studies and translational applications; and human islets are ultimately transplanted into patients.

As we will further elaborate in this chapter, the structural and functional features of pancreatic islets from different species should be carefully considered when optimizing conditions for islet isolation procedures to maximize islet yields and quality, and minimize their immunogenicity.

2.1.1 Cytoarchitecture

As already stated, pancreatic islets isolated from rodents (primarily mice) have been used extensively in islet research. Studies with mouse islets have provided a wealth of knowledge about the physiology and pathophysiology of the endocrine pancreas. Indeed, the mouse islet had been the prototypic pancreatic islet in textbooks and biomedical educational curricula. However, as availability of human pancreatic islets and their use in research became more common in the last two decades or so, indirect evidence about certain distinctive features of human islets had started to emerge [38–41]. But it was not until the middle of the last decade where two independent landmark studies, one by Cabrera and another by Brissova and their colleagues, have provided systematic experimental evidence on the unique structural and functional features of the human islet [42, 43]. These studies showed that the human pancreatic islet contains ≤50 % beta cells and ≥40 % alpha cells (Fig. 2.1a). This was in sharp contrast to the previously prevalent view of the pancreatic islet which was based on the mouse, where the beta cells, which are surrounded by a mantle of alpha and delta cells, typically account for up to 80 % of the islet (Fig. 2.1b).

Moreover, the evidence presented by the two studies by Cabrera et al. and Brissova et al. showed that the alpha, beta, and delta cells are intermingled throughout the human islet. The studies also showed that the intermingled cells were distributed along the blood vessels within the islet in no particular order [42, 43]. Moreover, Cabrera et al. showed that in human islets ≥90 % of the alpha and beta cells have heterotypic contacts with neighboring islet cells of another type. Based on this unique cytoarchitecture, it was proposed for the first time that the cellular arrangement in the human islet favored paracrine interactions among the different neighboring endocrine cells [43]. It was also suggested that the islet microcirculation did not necessarily dictate a specific hierarchical order within the human islet, where one endocrine cell may influence other downstreamcells during regulation islet function, as previously suggested for mouse islets [44–46]. Cabrera and colleagues further suggested that the intermingled distribution of the endocrine cells within the human islet reduced the electrical coupling between beta cells, which was in sharp contrast to what was previously reported for the mouse islet [47, 48]. Moreover, they showed reduced synchronization of cytoplasmic free calcium ($[Ca^{2+}]_i$) oscillations in beta cells throughout the whole human islet, as further evidence for diminished electrical coupling among the cells [43]. The association between the islet cytoarchitecture and synchronization of beta cell release during bursting activity in response to stimulus was further supported by a later study by Nittla and colleagues [49]. Together, these findings supported the notion that autocrine and paracrine signaling among the different endocrine cells in the human islet play significant roles in regulation of human islet function and overall glucose homeostasis [43, 50].

Nonhuman primates (monkeys) have been used as surrogates for human subjects in biomedical research for more than a century [51, 52]. Earlier comparative histopathological studies of the pancreas from different species including monkeys had shown different patterns of islet distribution and distinct arrangements of endocrine cells within the pancreatic islets [53]. Several later studies have shown that monkey pancreatic islets share many characteristics of the human islet (Fig. 2.1c) [42, 43, 50, 54, 55]. Monkey islets exhibit random distribution of endocrine cells along islet blood vessels with proportions of endocrine cells similar to those observed in human islets (see above) [43]. Importantly, much like human islets monkey islets have been shown to increase $[Ca^{2+}]_i$ signaling in response to lowering glucose, likely due to their higher proportion of alpha cells [43].

Pig islets are also used extensively in research. This has been motivated by the scarcity of human donor islets and the promise of unlimited availability of pig islets and other organs for xenotransplantation to respectively treat T1D and other organ-failure conditions in clinical applications [56, 57]. Although successful engraftment of pig islets after transplantation into monkeys has been shown, long lasting survival of the islet xenograft remains limited [58–60]. This is primarily due to strong immunity against tissues from other species which involve humoral, innate, and adaptive immune responses [61, 62]. However, with the advent of genome editing

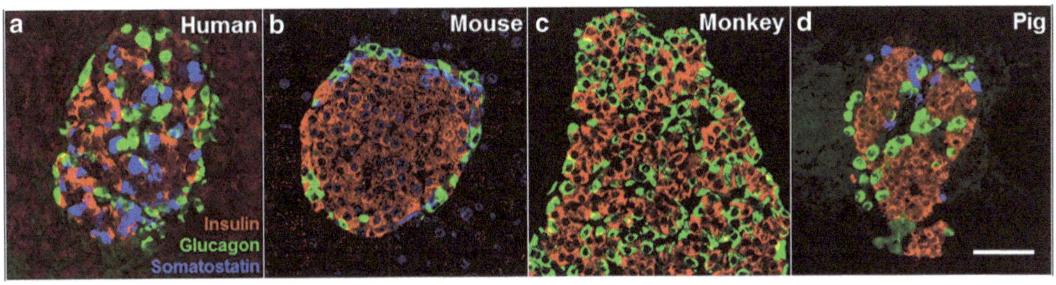

Fig. 2.1 Cytoarchitecture of the human, mouse, monkey, and pig islets. Immunostaining for insulin (*red*), glucagon (*green*), and somatostatin (*blue*) in fixed pancreatic sections obtained from (**a**) human, (**b**) mouse, (**c**) monkey, and (**d**) pig. Scale bar = 50 μm

techniques researchers have been able to modify/eliminate expression of certain pig antigens that have been known to be targets for anti-pig immunity in xenotransplantation [63]. While this process is expected to take some time before full fruition, where transplantation of pig pancreatic islets becomes standard-of-care in clinical therapy of T1D [64, 65], pig islets will continue to be isolated for general research purposes and preclinical applications.

Pig islets have been shown to share features of mouse islets where a single pig islet appears to be formed by a few smaller clusters resembling mouse islets (Fig. 2.1d) [43, 54]. Although it has been suggested that cellular composition and distribution of pig islets varies with age and location in the pancreas, the islet clusters are generally composed of a "core" of beta cells, accounting for ~90 % of the islet, which are surrounded mainly by alpha and delta cells [50, 66].

2.1.2 Vasculature

As mentioned above, the notion of hierarchical order of certain endocrine cells within the rodent pancreatic islet and the presumed consequences of this cellular organization on islet function have been prevalent in the literature. Much of this was primarily based on studies with rodent (mouse and rat) islets but only a handful of studies have indeed examined the dynamics of blood flow in the microcirculation of islets from other species [67, 68]. It had also been thought that the hierarchical order of endocrine cells is mediated through signaling from one islet part to another via blood flow in the microcirculation of the rodent islet [44, 46]. More recent in vivo studies have shown that two patterns of blood flow predominate in the mouse islet, where blood either first perfuses the core of the islet and flows outward toward the mantle or it flows from one side of the islet to the other in no particular direction and regardless of cell type [69]. Notably, based on some early ex vivo human pancreas perfusion studies and prevalent findings from rodent islets, it was also assumed that blood flow occurred from core to mantle in the human islet, despite the absence of anatomic or functional evidence to this effect, and that beta cell products controlled alpha and delta cell functions [45]. However, evidence presented in recent studies with human islets has indicated different signaling mechanisms among islet cells, not necessarily through blood flow. The evidence shows that paracrine signaling among the different endocrine cells, without a particular hierarchical order, is likely responsible for regulating the function of the human islet and overall glucose homeostasis [70–73].

Although the cytoarchitecture of the human islet and the intermingled arrangement of endocrine cells along islet capillaries did not support the notion of functional hierarchy based on blood flow alone [43], it did not necessarily exclude the possibility for additional layers of islet function regulation through blood flow. Blood flow can be regulated/modified by changes in blood vessel diameter [74, 75]. Changing vessel diameter, however, requires the presence of contractile elements in association with the islet macro and microvasculature to allow for vessel dilation/constriction and consequent changes in islet blood flow in response to functional demands on the endocrine pancreas [76]. Human pancreatic islets have been shown to contain abundant amounts of smooth muscle cells in association with blood vessels (Fig. 2.2) [77]. Pericytes were also shown to associate with a portion of the blood vessels in human islets. These findings indicate the presence of at least two populations of contractile cells in association with the microvasculature of the human islet. Notably, the abundant presence of contractile elements raises the possibility for localized regulation of blood flow within the human islet. Although a systematic characterization of the mechanisms underlying local regulation of blood flow within the human islet remains to be fully done, vasoactive compounds such as ATP and acetylcholine, which are released by endocrine cells in conjunction with hormones, may play a role in regulating blood flow locally and influencing the function of the human islet [70].

Another possibility for regulating the blood vessel diameter and blood flow in the human islet is through autonomic nervous input to the blood vessel-associated contractile elements, which are putative targets for autonomic sympathetic innervation. Indeed, we have shown that sympathetic nerve fibers primarily contact vascular structures within the human islet (Figs. 2.2 and 2.3a) [77].

In contrast to the human islet, capillaries in mouse islets are generally devoid of contractile elements except for one or two (depending on islet size) main arterioles, also known as feeding arterioles, which contain smooth muscle cells to change the diameter of the feeding arteriole(s) whereby regulating overall blood flow into the islet (Fig. 2.3b) [78]. This is further supported by a recent study showing that, irrespective of blood flow within the surrounding exocrine tissue, overall blood flow within the mouse islet is

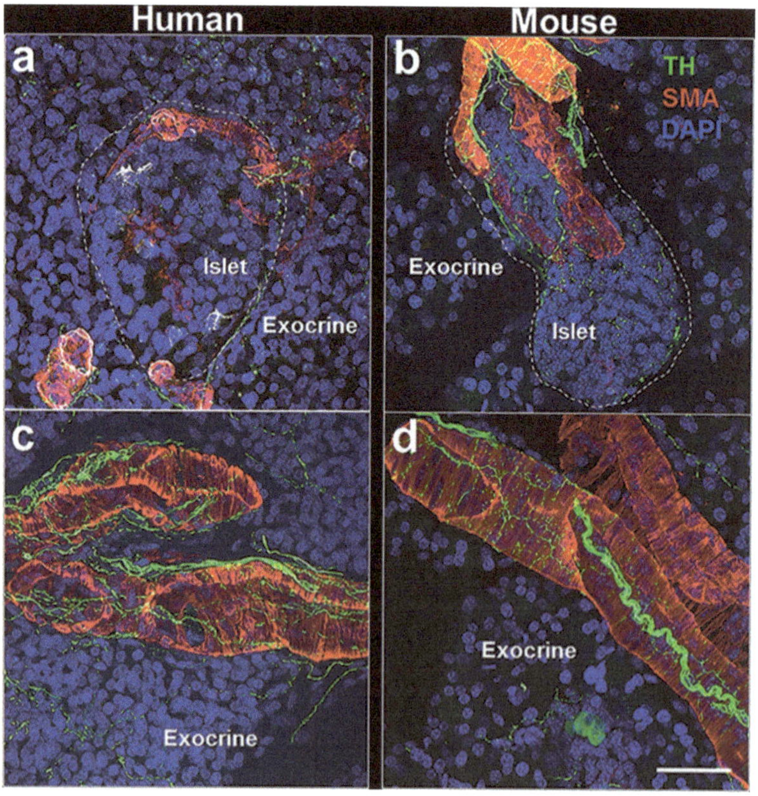

Fig. 2.2 Sympathetic innervation patterns in the human and mouse endocrine and exocrine pancreas. (**a**, **b**) Immunostaining for the sympathetic nerve marker tyrosine hydroxylase (TH; *green*) and smooth muscle actin (SMA; *red*) in the (**a**) human and (**b**) mouse endocrine pancreas (*blue*: DAPI nuclear counterstain). (**c**, **d**) Immunostaining for the same marker in (**c**) human and (**d**) and mouse exocrine tissue. Scale bar = 50 μm

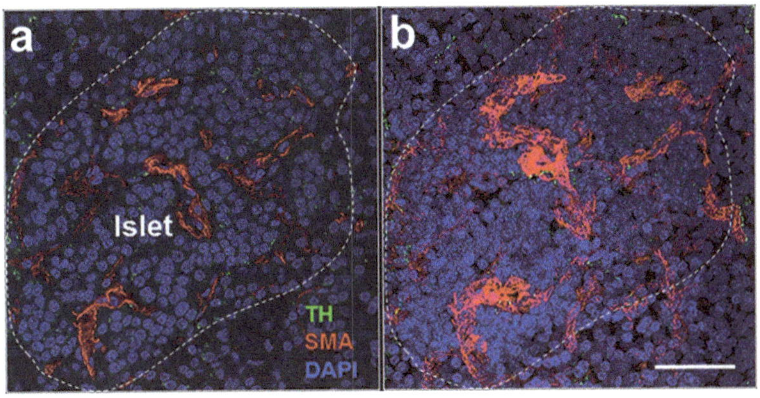

Fig. 2.3 Human islets have abundant smooth muscle actin in association with blood vessels. Immunostaining for smooth muscle actin (SMA; *red*) and the sympathetic nerve marker tyrosine hydroxylase (TH; *green*) in a human pancreatic islet shown in (**a**) a single confocal plane and (**b**) as maximum projection of a z-stack of multiple confocal images (*blue*: DAPI nuclear counterstain). Scale bar = 25 μm

uniformly regulated depending on islet functional demands in response to glucose metabolism [74].

Despite the above described differences in the vasculature of mouse and human islets, not all features of blood capillaries within human and mouse islets are different. Both human and mouse islets receive primary arterioles which branch into capillaries composed of single layer of endothelial cells [78]. On the ultrastructural level, both human and mouse islet capillaries show similar level of fenestration [79]. However, capillaries in mouse islets have a single basement membrane whereas those in human islets have been shown to be surrounded by a double membrane [55, 78, 80]. These ultra-structural differences and other contrasting vascular features among species should be taken into consideration during optimization of islet isolation procedures, as they may influence during enzymatic treatment the cellular integrity of islets and the overall quality of the isolation product.

2.1.3 Paracrine Signaling

Maintenance of glucose homeostasis requires intricate signaling and cross-talk among the different islet cells and complex coordination of the endocrine cell activity and their effects on peripheral target tissues (i.e., liver, muscle, and fat). It has been shown that hormones released by the different islet cells have different effects in the periphery. For example, beta cells release insulin in response to increased blood glucose levels (hyperglycemia) after food ingestion, where the released insulin promotes uptake of glucose from the circulation by the peripheral target tissues. Glucose update leads to lowering of plasma glucose levels (hypoglycemia) which in turn leads to alpha cells activation and release of glucagon. Glucagon release leads to an increase in blood sugar levels through neoglucogenesis and glycogenolysis primarily in the liver and muscles. Another example of this intricate signaling among endocrine cells is the delta cell release of somatostatin, which is thought to modulate secretion of both glucagon and insulin to avoid "overshooting" (i.e., excessive release). Therefore, paracrine signaling via islet hormones within the islet and onto peripheral target tissues is critical in the maintenance of glucose homeostasis.

The intricate regulation of glucose homeostasis involves additional signaling mechanisms other than hormones released by the islet endocrine cells. It has been shown that other factors released in conjunction with endocrine hormones (e.g., ATP, GABA, glutamate, acetylcholine, and Zn^{2+}), also have direct effects on the overall islet function through autocrine and paracrine signaling within the islet. We have shown that glutamate, released together with glucagon, acts as a positive autocrine factor on the alpha cell [81]. Glutamate primes the alpha cell and potentiates glucagon release in response to small fluctuations in blood glucose through ionotropic glutamate receptors of the AMPA/kainate type, which are expressed on alpha cells but not beta cells in the human islet. These findings highlight different functional aspect of the human islets in comparison to rodent islets where different effects of glutamate on alpha cell function have been suggested [82, 83].

GABA, another neurotransmitter co-released with insulin from beta cells, was also shown to inhibit glucagon secretion from alpha cells through paracrine effects mediated through $GABA_A$-receptor chloride channels [84]. Moreover, there have been conflicting reports on the paracrine effects of ATP on the islet function. ATP is released from beta cells in conjunction with insulin in response to glucose, and studies have shown both, excitatory and inhibitory effects of ATP on insulin release from mouse and rat islets [85–87]. In human islets, however, these effects have been reported to increase beta cell function and insulin release [88, 89]. It was also shown by Silva and colleagues that the ATP-gated purinergic receptor $P2X_3$ is the primary mediator of extracellular ATP action on the human beta cell [70]. Importantly, the authors showed that ATP has autocrine effects on the human beta cell leading to its sensitization and potentiation in response to subsequent stimulation by glucose. Moreover, Silva et al. suggested that ATP may also be released from insulin granules, at low glucose concentration not sufficient to induce insulin likely via the previously described kiss-and-run membrane fusion events, to prime the beta cells for robust insulin release upon subsequent stimulation by high glucose [90, 91].

The delta cells, the third major cell type of the endocrine pancreas, are thought to influence glucose metabolism through release of somatostatin and its inhibitory effects on insulin and glucagon secretion. The regulatory mechanisms of the human delta cell activity have been scarcely investigated despite the purported influence of the delta cell on overall islet function. In contrast to alpha and beta cells, little is known about the signaling mechanisms that regulate somatostatin secretion from delta cells in the human pancreatic islet. This is critical information because of the putative inhibitory effects of somatostatin on insulin and glucagon release and the sparse distribution of delta cells within the human islet [81]. While most signaling molecules present and released in the human islet have not been thoroughly evaluated for their effects on the human delta cell, GABA was recently shown to elicit large depolarizing currents in the delta cell [92]. We have also recently shown that delta cells in human islets receive the highest density of sympathetic innervation [77]. Together, these findings suggest that the function of the delta cell in the human islet may be regulated by paracrine signaling mechanisms as well as autonomic nerve input.

2.1.4 Innervation

Earlier indications about potential influence of the nervous system on the regulation of blood sugar levels were revealed by Claude Bernard in the mid 1800s [93]. It was later shown in the 1900s by Paul Langerhans that the pancreatic islet is richly innervated [94]. Since then, many studies have implicated the autonomic nervous system in islet function regulation and overall glucose homeostasis [95–100], but the role of direct autonomic input to the islet remains poorly understood because of the many mechanisms the autonomic nervous system may use, within the islet and the periphery, to influence overall glucose homeostasis.

Although normal blood sugar values do vary among species, many animals including mammals are capable of maintaining strikingly narrow ranges of blood sugar levels under normal conditions. Such stringent and critical control of glycemia likely requires various and complex mechanisms. We discussed above some of the autocrine and paracrine signaling mechanisms within the endocrine pancreas and in its peripheral target organs (e.g., liver and muscles). Here, we further discuss the role of the autonomic nervous system in regulating and modulating glucose homeostasis.

The autonomic nervous system consists of two arms, sympathetic and parasympathetic, which innervate vital organs and have critical roles in maintaining overall organism homeostasis, including blood sugar control. In doing so, it ensures availability of glucose as fuel for vital functions (e.g., brain function) under different physiological states and environmental conditions (e.g., digestion, fight-or-flight response, hypothermia, etc.) [101–104]. In the endocrine pancreas, the sympathetic input to the pancreatic islet stimulates glucagon release and inhibits insulin secretion, and the parasympathetic system is thought to stimulate release of both insulin and glucagon [97]. Thus, the autonomic nervous system plays a critical role in regulating the function of the endocrine pancreas and overall glucose homeostasis [95, 105]. However, evidence from transplantation studies of whole pancreas or isolated islets, where "direct" innervation of the endocrine pancreas is absent, suggests the presence of compensatory mechanism to help maintain glucose homeostasis.

Nonetheless, it is well established that the pancreas, endocrine and exocrine, receives sympathetic, parasympathetic, and even sensory nervous input (Fig. 2.3). Evidence indicates that the innervation patterns in the mouse and human exocrine pancreas are similar (Figs. 2.3c, d). Therefore, the same has been assumed for the endocrine pancreas, and that neuronal modulation of the function in the human endocrine pancreas is mediated by direct innervation of the pancreatic islet and its cells [106–108]. However, most of the studies of islet innervation have been conducted with rodent islets and there have been only a few non-comprehensive studies with human islets to convincingly conclude similar innervation patterns to rodent islets [109–111]. Importantly, human islets have not been examined for the presence of classical markers of the parasympathetic and sympathetic systems [112, 113]. Furthermore, the exact location(s) where neuronal axons terminate within the human islets were not shown until recently [77].

The findings of Rodriguez-Diaz and colleagues revealed differences in the innervation patterns of the autonomic nervous system in mouse and human islets [77]. They showed that mouse islets are densely innervated by sympathetic and parasympathetic nerve fibers where the former primarily contact alpha and delta cells and the latter alpha and beta cells (Figs. 2.2b and 2.4b) [105, 114]. In contrast, the authors showed that human islets have fewer sympathetic nerve fibers which are found along blood vessels with preferential localization near the vessel-associated contractile elements, and far fewer parasympathetic fibers compared to mouse islets (Figs. 2.2, 2.3a, and 2.4a). Instead, they showed that parasympathetic effects in the human islet are likely mediated through release of acetylcholine, a major parasympatheticneurotransmitter, from the alpha cells (Fig. 2.4a) [71].

In summary, while the autonomic nervous system plays a critical role in modulating the pancreatic islet function and overall glucose homeostasis, it does so using different mechanisms in rodent and human islets. In mice, the autonomic nerve input to the endocrine pancreas is likely mediated through direct contact with the islet cells [105, 114]. In humans, however, the effects of the sympathetic innervation are likely mediated through indirect effects on local blood flow within the islet microcirculation. Given the scarce presence of parasympathetic nerve fibers in the human islet, it is likely that direct parasympathetic influence on islet function is significantly less in humans compared to mice [71, 77].

2.2 Summary

In this chapter, we have explored distinct structural and functional features of pancreatic islets from different species. It is evident that the pancreatic islet has different faces (i.e., features) which should be considered when optimizing isolation protocols and procedures. This will help maximize islet yields and quality in pre-clinical and clinical applications. Importantly, optimal isolation conditions will minimize the immunogenicity of pancreatic islets isolated for the purpose of transplantation into human patients. This will ultimately help improve islet graft survival and clinical outcome in islet transplantation as beta cell replacement therapy of T1D.

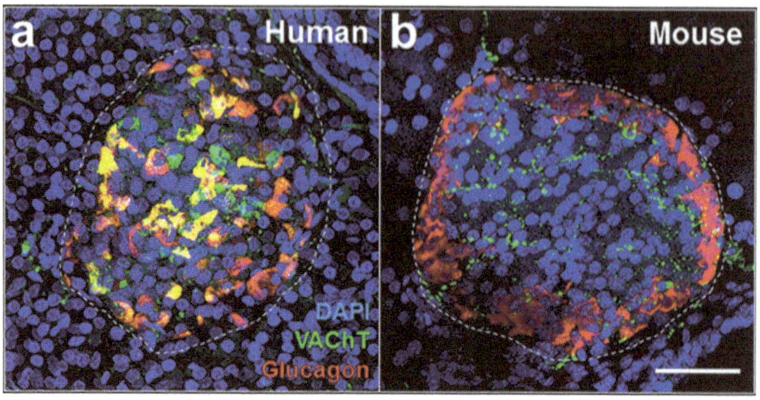

Fig. 2.4 Human and mouse pancreatic islets have different patterns of parasympathetic innervation. Immunostaining for the parasympathetic nerve marker vesicular acetylcholine esterase (VAchT; *green*) and glucagon (*red*) in a (**a**) human and (**b**) mouse islet (*blue*: DAPI nuclear counterstain). Scale bar = 50 μm

References

1. Grauslund J. Eye complications and markers of morbidity and mortality in long-term type 1 diabetes. Acta Ophthalmol. 2011;89(Thesis 1):1–19.
2. Scheen AJ. Pharmacokinetics and clinical use of incretin-based therapies in patients with chronic kidney disease and type 2 diabetes. Clin Pharmacokinet. 2014;53:773–85.
3. Narayan KM, Boyle JP, Geiss LS, Saaddine JB, Thompson TJ. Impact of recent increase in incidence on future diabetes burden: U.S., 2005–2050. Diabetes Care. 2006;29(9):2114–6.
4. Pugliese A. The multiple origins of type 1 diabetes. Diabet Med. 2013;30(2):135–46.
5. Abdulreda MH, Berggren PO. Islet inflammation in plain sight. Diabetes Obes Metab. 2013;15 Suppl 3:105–16.
6. Hokanson JE, Kinney GL, Cheng S, Erlich HA, Kretowski A, Rewers M. Susceptibility to type 1 diabetes is associated with ApoCIII gene haplotypes. Diabetes. 2006;55(3):834–8.
7. de Luca C, Olefsky JM. Inflammation and insulin resistance. FEBS Lett. 2008;582(1):97–105.
8. Shoelson SE, Lee J, Goldfine AB. Inflammation and insulin resistance. J Clin Invest. 2006;116(7):1793–801.
9. Bruun JM, Helge JW, Richelsen B, Stallknecht B. Diet and exercise reduce low-grade inflammation and macrophage infiltration in adipose tissue but not in skeletal muscle in severely obese subjects. Am J Physiol Endocrinol Metab. 2006;290(5):E961–7.
10. Kim D, Kim J, Yoon JH, Ghim J, Yea K, Song P, et al. CXCL12 secreted from adipose tissue recruits macrophages and induces insulin resistance in mice. Diabetologia. 2014;57(7):1456–65.
11. Olivieri O, Martinelli N, Bassi A, Trabetti E, Girelli D, Pizzolo F, et al. ApoE epsilon2/epsilon3/epsilon4 polymorphism, ApoC-III/ApoE ratio and metabolic syndrome. Clin Exp Med. 2007;7(4):164–72.
12. Caron S, Verrijken A, Mertens I, Samanez CH, Mautino G, Haas JT, et al. Transcriptional activation of apolipoprotein CIII expression by glucose may contribute to diabetic dyslipidemia. Arterioscler Thromb Vasc Biol. 2011;31(3):513–9.
13. Oluleye TS. Diabetic retinopathy: current developments in pathogenesis and management. Afr J Med Med Sci. 2010;39(3):199–206.
14. Eguchi K, Manabe I, Oishi-Tanaka Y, Ohsugi M, Kono N, Ogata F, et al. Saturated fatty acid and TLR signaling link β cell dysfunction and islet inflammation. Cell Metab. 2012;15(4):518–33.
15. Timper K, Donath MY. Diabetes mellitus type 2 – the new face of an old lady. Swiss Med Wkly. 2012;142:w13635.
16. Staels W, Heremans Y, Heimberg H. Reprogramming of human exocrine pancreas cells to beta cells. Best Pract Res Clin Endocrinol Metab. 2015;29(6):849–57.
17. Stanojevic V, Habener JF. Evolving function and potential of pancreatic alpha cells. Best Pract Res Clin Endocrinol Metab. 2015;29(6):859–71.
18. Mehrfarjam Z, Esmaeili F, Shabani L, Ebrahimie E. Induction of pancreatic beta cell gene expression in mesenchymal stem cells. Cell Biol Int. 2016;40(5):486–500.
19. Assouline-Thomas B, Ellis D, Petropavlovskaia M, Makhlin J, Ding J, Rosenberg L. Islet Neogenesis Associated Protein (INGAP) induces the differentiation of an adult human pancreatic ductal cell line into insulin-expressing cells through stepwise activation of key transcription factors for embryonic beta cell development. Differentiation. 2015;90:77–90.

20. Shapiro A, Ricordi C, Hering B, Auchincloss H, Lindblad R, Robertson R, et al. International trial of the Edmonton protocol for islet transplantation. N Engl J Med. 2006;355(13):1318–30.
21. Froud T, Ricordi C, Baidal DA, Hafiz MM, Ponte G, Cure P, et al. Islet transplantation in type 1 diabetes mellitus using cultured islets and steroid-free immunosuppression: Miami experience. Am J Transplant. 2005;5(8):2037–46.
22. Shapiro AM, Lakey JR, Ryan EA, Korbutt GS, Toth E, Warnock GL, et al. Islet transplantation in seven patients with type 1 diabetes mellitus using a glucocorticoid-free immunosuppressive regimen. N Engl J Med. 2000;343(4):230–8.
23. Hogan A, Pileggi A, Ricordi C. Transplantation: current developments and future directions; the future of clinical islet transplantation as a cure for diabetes. Front Biosci. 2008;13:1192–205.
24. Ricordi C, Strom TB. Clinical islet transplantation: advances and immunological challenges. Nat Rev Immunol. 2004;4(4):259–68.
25. Pileggi A, Ricordi C, Kenyon NS, Froud T, Baidal DA, Kahn A, et al. Twenty years of clinical islet transplantation at the Diabetes Research Institute – University of Miami. Clin Transplant. 2004:177–204.
26. Johansson H, Lukinius A, Moberg L, Lundgren T, Berne C, Foss A, et al. Tissue factor produced by the endocrine cells of the islets of Langerhans is associated with a negative outcome of clinical islet transplantation. Diabetes. 2005;54(6):1755–62.
27. Moberg L, Johansson H, Lukinius A, Berne C, Foss A, Kallen R, et al. Production of tissue factor by pancreatic islet cells as a trigger of detrimental thrombotic reactions in clinical islet transplantation. Lancet. 2002;360(9350):2039–45.
28. Loganathan G, Dawra RK, Pugazhenthi S, Guo Z, Soltani SM, Wiseman A, et al. Insulin degradation by acinar cell proteases creates a dysfunctional environment for human islets before/after transplantation: benefits of α-1 antitrypsin treatment. Transplantation. 2011;92(11):1222–30.
29. Linetsky E, Inverardi L, Kenyon NS, Alejandro R, Ricordi C. Endotoxin contamination of reagents used during isolation and purification of human pancreatic islets. Transplant Proc. 1998;30(2):345–6.
30. Inverardi L, Linetsky E, Kenyon NS, Socci C, Ricordi C. Human mixed lymphocyte-islet cultures: the influence of heterologous proteins on islet immunogenicity. Transplant Proc. 1997;29(4):2066.
31. Balamurugan AN, He J, Guo F, Stolz DB, Bertera S, Geng X, et al. Harmful delayed effects of exogenous isolation enzymes on isolated human islets: relevance to clinical transplantation. Am J Transplant. 2005;5(11):2671–81.
32. Balamurugan AN, Naziruddin B, Lockridge A, Tiwari M, Loganathan G, Takita M, et al. Islet product characteristics and factors related to successful human islet transplantation from the Collaborative Islet Transplant Registry (CITR) 1999–2010. Am J Transplant. 2014;14(11):2595–606.
33. Mita A, Ricordi C, Messinger S, Miki A, Misawa R, Barker S, et al. Antiproinflammatory effects of iodixanol (OptiPrep)-based density gradient purification on human islet preparations. Cell Transplant. 2010;19(12):1537–46.
34. Park SG, Kim JH, Oh JH, Lee HN, Park HS, Chung SS, et al. Polymyxin B, scavenger of endotoxin, enhances isolation yield and in vivo function of islets. Transpl Int. 2010;23(3):325–32.
35. Fernandez L, Lehmann R, Selvaggi G, Kong SS, Bottino R, Li H, et al. Influence of variables on canine islet isolation results. Transplant Proc. 1997;29(4):1950.
36. Linetsky E, Bottino R, Lehmann R, Alejandro R, Inverardi L, Ricordi C. Improved human islet isolation using a new enzyme blend, liberase. Diabetes. 1997;46(7):1120–3.
37. Itoh T, Sugimoto K, Shimoda M, Chujo D, Takita M, Iwahashi S, et al. Establishment of a prolonged pancreas preservation model for islet isolation research in mice. Islets. 2011;3(6):376–80.
38. Rahier J, Goebbels RM, Henquin JC. Cellular composition of the human diabetic pancreas. Diabetologia. 1983;24(5):366–71.
39. Stefan Y, Orci L, Malaisse-Lagae F, Perrelet A, Patel Y, Unger RH. Quantitation of endocrine cell content in the pancreas of nondiabetic and diabetic humans. Diabetes. 1982;31(8 Pt 1):694–700.
40. Clark A, Wells CA, Buley ID, Cruickshank JK, Vanhegan RI, Matthews DR, et al. Islet amyloid, increased A-cells, reduced B-cells and exocrine fibrosis: quantitative changes in the pancreas in type 2 diabetes. Diabetes Res. 1988;9(4):151–9.
41. Butler AE, Janson J, Bonner-Weir S, Ritzel R, Rizza RA, Butler PC. Beta-cell deficit and increased beta-cell apoptosis in humans with type 2 diabetes. Diabetes. 2003;52(1):102–10.
42. Brissova M, Fowler M, Nicholson W, Chu A, Hirshberg B, Harlan D, et al. Assessment of human pancreatic islet architecture and composition by laser scanning confocal microscopy. J Histochem Cytochem. 2005;53(9):1087–97.
43. Cabrera O, Berman DM, Kenyon NS, Ricordi C, Berggren PO, Caicedo A. The unique cytoarchitecture of human pancreatic islets has implications for islet cell function. Proc Natl Acad Sci U S A. 2006;103(7):2334–9.
44. Bonner-Weir S, Orci L. New perspectives on the microvasculature of the islets of Langerhans in the rat. Diabetes. 1982;31(10):883–9.
45. Stagner JI, Samols E. The vascular order of islet cellular perfusion in the human pancreas. Diabetes. 1992;41(1):93–7.
46. Brunicardi F, Stagner J, Bonner-Weir S, Wayland H, Kleinman R, Livingston E, et al. Microcirculation of the islets of Langerhans. Long Beach Veterans Administration Regional Medical Education Center Symposium. Diabetes. 1996;45(4):385–92.
47. Ravier M, Güldenagel M, Charollais A, Gjinovci A, Caille D, Söhl G, et al. Loss of connexin36 channels

alters beta-cell coupling, islet synchronization of glucose-induced Ca2+ and insulin oscillations, and basal insulin release. Diabetes. 2005;54(6): 1798–807.
48. Serre-Beinier V, Bosco D, Zulianello L, Charollais A, Caille D, Charpantier E, et al. Cx36 makes channels coupling human pancreatic beta-cells, and correlates with insulin expression. Hum Mol Genet. 2009;18(3):428–39.
49. Nittala A, Ghosh S, Wang X. Investigating the role of islet cytoarchitecture in its oscillation using a new beta-cell cluster model. PLoS ONE. 2007;2(10), e983.
50. Kim A, Miller K, Jo J, Kilimnik G, Wojcik P, Hara M. Islet architecture: a comparative study. Islets. 2009;1(2):129–36.
51. Meltzer SJ, Auer J. The effects of intraspinal injection of magnesium salts upon tetanus. J Exp Med. 1906;8(6):692–706.
52. Flexner S. Experimental cerebro-spinal meningitis in monkeys. J Exp Med. 1907;9(2):142–67.
53. Wieczorek G, Pospischil A, Perentes E. A comparative immunohistochemical study of pancreatic islets in laboratory animals (rats, dogs, minipigs, nonhuman primates). Exp Toxicol Pathol. 1998;50(3): 151–72.
54. Steiner DJ, Kim A, Miller K, Hara M. Pancreatic islet plasticity: interspecies comparison of islet architecture and composition. Islets. 2010;2(3): 135–45.
55. Arrojo e Drigo R, Ali Y, Diez J, Srinivasan DK, Berggren PO, Boehm BO. New insights into the architecture of the islet of Langerhans: a focused cross-species assessment. Diabetologia. 2015;58(10):2218–28.
56. Vadori M, Cozzi E. The immunological barriers to xenotransplantation. Tissue Antigens. 2015;86(4): 239–53.
57. Cooper DK, Bottino R. Recent advances in understanding xenotransplantation: implications for the clinic. Expert Rev Clin Immunol. 2015;11(12):1379–90.
58. Shin JS, Kim JM, Kim JS, Min BH, Kim YH, Kim HJ, et al. Long-term control of diabetes in immunosuppressed Nonhuman Primates (NHP) by the transplantation of adult porcine islets. Am J Transplant. 2015;15(11):2837–50.
59. Bottino R, Wijkstrom M, van der Windt DJ, Hara H, Ezzelarab M, Murase N, et al. Pig-to-monkey islet xenotransplantation using multi-transgenic pigs. Am J Transplant. 2014;14(10):2275–87.
60. van der Windt DJ, Bottino R, Casu A, Campanile N, Cooper DK. Rapid loss of intraportally transplanted islets: an overview of pathophysiology and preventive strategies. Xenotransplantation. 2007;14(4):288–97.
61. Byrne GW, McGregor CG, Breimer ME. Recent investigations into pig antigen and anti-pig antibody expression. Int J Surg. 2015;23:223–8.
62. Kang HJ, Lee H, Park EM, Kim JM, Shin JS, Kim JS, et al. Increase in anti-Gal IgM level is associated with early graft failure in intraportal porcine islet xenotransplantation. Ann Lab Med. 2015;35(6): 611–7.
63. Estrada JL, Martens G, Li P, Adams A, Newell KA, Ford ML, et al. Evaluation of human and nonhumanprimate antibody binding to pig cells lacking GGTA1/CMAH/beta4GalNT2 genes. Xenotransplantation. 2015;22(3):194–202.
64. Ellis CE, Korbutt GS. Justifying clinical trials for porcine islet xenotransplantation. Xenotransplantation. 2015;22(5):336–44.
65. Schuurman HJ. Regulatory aspects of clinical xenotransplantation. Int J Surg. 2015;23:312–21.
66. Jay TR, Heald KA, Carless NJ, Topham DE, Downing R. The distribution of porcine pancreatic beta-cells at ages 5, 12 and 24 weeks. Xenotransplantation. 1999;6(2):131–40.
67. Stagner JI, Samols E, Koerker DJ, Goodner CJ. Perfusion with anti-insulin gamma globulin indicates a B to A to D cellular perfusion sequence in the pancreas of the rhesus monkey, Macaca mulatta. Pancreas. 1992;7(1):26–9.
68. Stagner JI, Samols E, Bonner-Weir S. Beta – alpha – delta pancreatic islet cellular perfusion in dogs. Diabetes. 1988;37(12):1715–21.
69. Nyman LR, Wells KS, Head WS, McCaughey M, Ford E, Brissova M, et al. Real-time, multidimensional in vivo imaging used to investigate blood flow in mouse pancreatic islets. J Clin Invest. 2008;118(11):3790–7.
70. Jacques-Silva MC, Correa-Medina M, Cabrera O, Rodriguez-Diaz R, Makeeva N, Fachado A, et al. ATP-gated P2X3 receptors constitute a positive autocrine signal for insulin release in the human pancreatic beta cell. Proc Natl Acad Sci U S A. 2010;107(14):6465–70.
71. Rodriguez-Diaz R, Dando R, Jacques-Silva MC, Fachado A, Molina J, Abdulreda MH, et al. Alpha cells secrete acetylcholine as a non-neuronal paracrine signal priming beta cell function in humans. Nat Med. 2011;17(7):888–92.
72. Rodriguez-Diaz R, Menegaz D, Caicedo A. Neurotransmitters act as paracrine signals to regulate insulin secretion from the human pancreatic islet. J Physiol. 2014;592(Pt 16):3413–7.
73. Rodriguez-Diaz R, Dando R, Huang YA, Berggren PO, Roper SD, Caicedo A. Real-time detection of acetylcholine release from the human endocrine pancreas. Nat Protoc. 2012;7(6):1015–23.
74. Nyman LR, Ford E, Powers AC, Piston DW. Glucose-dependent blood flow dynamics in murine pancreatic islets in vivo. Am J Physiol Endocrinol Metab. 2010;298(4):E807–14.
75. Dai C, Brissova M, Reinert RB, Nyman L, Liu EH, Thompson C, et al. Pancreatic islet vasculature adapts to insulin resistance through dilation and not angiogenesis. Diabetes. 2013;62(12):4144–53.
76. Ballian N, Brunicardi FC. Islet vasculature as a regulator of endocrine pancreas function. World J Surg. 2007;31(4):705–14.

77. Rodriguez-Diaz R, Abdulreda MH, Formoso AL, Gans I, Ricordi C, Berggren PO, et al. Innervation patterns of autonomic axons in the human endocrine pancreas. Cell Metab. 2011;14(1):45–54.
78. Dolensek J, Rupnik MS, Stozer A. Structural similarities and differences between the human and the mouse pancreas. Islets. 2015;7(1), e1024405.
79. Nyqvist D, Speier S, Rodriguez-Diaz R, Molano RD, Lipovsek S, Rupnik M, et al. Donor islet endothelial cells in pancreatic islet revascularization. Diabetes. 2011;60(10):2571–7.
80. Virtanen I, Banerjee M, Palgi J, Korsgren O, Lukinius A, Thornell LE, et al. Blood vessels of human islets of Langerhans are surrounded by a double basement membrane. Diabetologia. 2008;51(7):1181–91.
81. Cabrera O, Jacques-Silva MC, Speier S, Yang SN, Köhler M, Fachado A, et al. Glutamate is a positive autocrine signal for glucagon release. Cell Metab. 2008;7(6):545–54.
82. Moriyama Y, Hayashi M. Glutamate-mediated signaling in the islets of Langerhans: a thread entangled. Trends Pharmacol Sci. 2003;24(10):511–7.
83. Bertrand G, Gross R, Puech R, Loubatières-Mariani M, Bockaert J. Glutamate stimulates glucagon secretion via an excitatory amino acid receptor of the AMPA subtype in rat pancreas. Eur J Pharmacol. 1993;237(1):45–50.
84. Rorsman P, Berggren PO, Bokvist K, Ericson H, Möhler H, Ostenson CG, et al. Glucose-inhibition of glucagon secretion involves activation of GABAA-receptor chloride channels. Nature. 1989;341(6239):233–6.
85. Petit P, Hillaire-Buys D, Manteghetti M, Debrus S, Chapal J, Loubatières-Mariani M. Evidence for two different types of P2 receptors stimulating insulin secretion from pancreatic B cell. Br J Pharmacol. 1998;125(6):1368–74.
86. Salehi A, Qader S, Quader S, Grapengiesser E, Hellman B. Inhibition of purinoceptors amplifies glucose-stimulated insulin release with removal of its pulsatility. Diabetes. 2005;54(7):2126–31.
87. Léon C, Freund M, Latchoumanin O, Farret A, Petit P, Cazenave J, et al. The P2Y(1) receptor is involved in the maintenance of glucose homeostasis and in insulin secretion in mice. Purinergic Signal. 2005;1(2):145–51.
88. Fernandez-Alvarez J, Hillaire-Buys D, Loubatières-Mariani M, Gomis R, Petit P. P2 receptor agonists stimulate insulin release from human pancreatic islets. Pancreas. 2001;22(1):69–71.
89. Silva A, Rodrigues R, Tomé A, Cunha R, Misler S, Rosário L, et al. Electrophysiological and immunocytochemical evidence for P2X purinergic receptors in pancreatic beta cells. Pancreas. 2008;36(3):279–83.
90. Hanna S, Pigeau G, Galvanovskis J, Clark A, Rorsman P, MacDonald P. Kiss-and-run exocytosis and fusion pores of secretory vesicles in human beta-cells. Pflugers Arch. 2009;457(6):1343–50.
91. MacDonald P, Braun M, Galvanovskis J, Rorsman P. Release of small transmitters through kiss-and-run fusion pores in rat pancreatic beta cells. Cell Metab. 2006;4(4):283–90.
92. Braun M, Ramracheya R, Bengtsson M, Clark A, Walker JN, Johnson PR, et al. Gamma-aminobutyric acid (GABA) is an autocrine excitatory transmitter in human pancreatic beta-cells. Diabetes. 2010;59:1694–701. United States.
93. Bernard C. Leçons de physiologie expérimentale appliquée à la médecine, faites au Collège de France, par m. Claude Bernard. Paris: J.B. Baillière et fils; 1855 [etc., etc.].
94. Langerhans P, Morrison H. Contributions to the microscopic anatomy of the pancreas. Baltimore: The Johns Hopkins Press; 1937. 1 p.l., 39 p. p.
95. Woods S, Porte DJ. Neural control of the endocrine pancreas. Physiol Rev. 1974;54(3):596–619.
96. Satin L, Kinard T. Neurotransmitters and their receptors in the islets of Langerhans of the pancreas: what messages do acetylcholine, glutamate, and GABA transmit? Endocrine. 1998;8(3):213–23.
97. Ahrén B. Autonomic regulation of islet hormone secretion – implications for health and disease. Diabetologia. 2000;43(4):393–410.
98. Gilon P, Henquin J. Mechanisms and physiological significance of the cholinergic control of pancreatic beta-cell function. Endocr Rev. 2001;22(5):565–604.
99. Brunicardi F, Shavelle D, Andersen D. Neural regulation of the endocrine pancreas. Int J Pancreatol. 1995;18(3):177–95.
100. Havel P, Taborsky GJ. The contribution of the autonomic nervous system to changes of glucagon and insulin secretion during hypoglycemic stress. Endocr Rev. 1989;10(3):332–50.
101. D'Alessio D, Kieffer T, Taborsky GJ, Havel P. Activation of the parasympathetic nervous system is necessary for normal meal-induced insulin secretion in rhesus macaques. J Clin Endocrinol Metab. 2001;86(3):1253–9.
102. Ahrén B, Holst J. The cephalic insulin response to meal ingestion in humans is dependent on both cholinergic and noncholinergic mechanisms and is important for postprandial glycemia. Diabetes. 2001;50(5):1030–8.
103. Teff KL, Mattes RD, Engelman K. Cephalic phase insulin release in normal weight males: verification and reliability. Am J Physiol. 1991;261(4 Pt 1): E430–6.
104. Chenon D, Ribes G, Loubatières-Mariani M. Importance of the cholinergic nervous system in the postprandial secretion of insulin in dogs. C R Seances Soc Biol Fil. 1984;178(3):307–12.
105. Rodriguez-Diaz R, Caicedo A. Novel approaches to studying the role of innervation in the biology of pancreatic islets. Endocrinol Metab Clin N Am. 2013;42(1):39–56.
106. Lindskog S, Ahrén B, Dunning B, Sundler F. Galanin-immunoreactive nerves in the mouse and rat pancreas. Cell Tissue Res. 1991;264(2):363–8.

107. Karlsson S, Sundler F, Ahrén B. Neonatal capsaicin-treatment in mice: effects on pancreatic peptidergic nerves and 2-deoxy-D-glucose-induced insulin and glucagon secretion. J Auton Nerv Syst. 1992;39(1):51–9.
108. Gautam D, Jeon J, Li J, Han S, Hamdan F, Cui Y, et al. Metabolic roles of the M3 muscarinic acetylcholine receptor studied with M3 receptor mutant mice: a review. J Recept Signal Transduct Res. 2008;28(1–2):93–108.
109. Ahrén B, Ar'Rajab A, Böttcher G, Sundler F, Dunning B. Presence of galanin in human pancreatic nerves and inhibition of insulin secretion from isolated human islets. Cell Tissue Res. 1991;264(2):263–7.
110. Bishop AE, Polak JM, Green IC, Bryant MG, Bloom SR. The location of VIP in the pancreas of man and rat. Diabetologia. 1980;18(1):73–8.
111. Amenta F, Cavallotti C, de Rossi M, Tonelli F, Vatrella F. The cholinergic innervation of human pancreatic islets. Acta Histochem. 1983;73(2):273–8.
112. Ding WG, Kimura H, Fujimura M, Fujimiya M. Neuropeptide Y and peptide YY immunoreactivities in the pancreas of various vertebrates. Peptides. 1997;18(10):1523–9.
113. Shimosegawa T, Asakura T, Kashimura J, Yoshida K, Meguro T, Koizumi M, et al. Neurons containing gastrin releasing peptide-like immunoreactivity in the human pancreas. Pancreas. 1993;8(4):403–12.
114. Rodriguez-Diaz R, Speier S, Molano RD, Formoso A, Gans I, Abdulreda MH, et al. Noninvasive in vivo model demonstrating the effects of autonomic innervation on pancreatic islet function. Proc Natl Acad Sci U S A. 2012;109(52):21456–61.

Isolation of Mouse Pancreatic Islets of Langerhans

3

Miriam Ramírez-Domínguez

Abstract

The aim of any pancreatic islet isolation is obtaining pure, viable and functional pancreatic islets, either for in vitro or in vivo purposes. The islets of Langerhans are complex microorgans with the important role of regulating glucose homeostasis. Imbalances in glucose homeostasis lead to diabetes, which is defined by the American Diabetes Association as a "group of metabolic diseases characterized by hyperglycemia resulting from defects in insulin secretion, insulin action or both" (American Diabetes Association 2011). Currently, the rising demand of human islets is provoking a shortage of this tissue, limiting research and clinical practice on this field. In this scenario, it is essential to investigate and improve islet isolation procedures in animal models, while keeping in mind the anatomical and functional differences between species. This chapter discusses the main aspects of mouse islet isolation research, highlighting the critical factors and shortcomings to take into account for the selection and/or optimization of a mouse islet isolation protocol.

Keywords

Diabetes • Islet isolation • Islets of Langerhans • Islet transplantation • Islet purification • Animal models • Translational research

3.1 Introduction

Modern islet research started with Bensley and his pioneering scientific contribution in 1911 [2]. One hundred years later, the field has evolved tremendously, especially since the success of the Edmonton protocol. It was developed by the islet transplantation team at the University of Alberta, in Canada, which first introduced a steroid-free immunosup-

M. Ramírez-Domínguez (✉)
Laboratory of Cell Therapy of Diabetes, Department of Pediatrics, Faculty of Medicine and Odontology, Hospital Cruces, University of the Basque Country (UPV/EHU), Barrio Sarriena, s/n 48940, Leioa, Biscay, Spain
e-mail: miriamrd@gmail.com

M. Ramírez-Domínguez (ed.), *Pancreatic Islet Isolation*, Advances in Experimental Medicine and Biology 938, DOI 10.1007/978-3-319-39824-2_3

pressive regimen, resulting in 100% insulin independence at 1 year in seven individuals [3]. This advance contributed to the worldwide expansion of human transplantation program and the access to human tissue for translational studies.

The islet community has recently appealed for a higher investment in human islet isolation and distribution to the NIH, JDRF and American Diabetes Association [4, 5]. The current rising number of researchers working with human islets [from 35 active users in the Integrated Islet Distribution Program (IIDP) in 2010 to 104 in 2014] and the consequent rising demand for this tissue has resulted in a bottleneck in the research islet supply. Thus, the use of mice as animal models for islet isolation and other in vitro and in vivo purposes has emerged as an alternative to study the pathophysiology of diabetes as well as to conduct islet isolation and transplantation.

The aim of pancreatic islet isolation is obtaining viable, pure and functional islets, either for in vitro or in vivo studies (Fig. 3.1). However, to obtain a successful yield and good quality islets, different key aspects must be taken into account: the type and concentration of the digestive enzyme, the method of enzyme administration, the temperature and duration of the pancreas digestion, the method for islet purification and the culture conditions following isolation [6]. It is instrumental to identify the factors influencing the efficacy of the isolation procedure of mouse pancreatic islets in order to standardize the procedure, reduce the variability, and harvest good quality islets.

The main steps in any mouse islet isolation are the following: pancreas distention and dissection; pancreas digestion; islet purification and islet culture. In this chapter, we describe the main methodological aspects of mouse islet isolation as well as the key aspects and challenges.

3.2 Pancreas Distention and Dissection

3.2.1 General Considerations About the Procedure

First of all, the animal should be euthanized by cervical dislocation, CO_2 asphyxiation, etc., depending on the regulations in that country and/or the laboratory choice. While it is desirable to perform all the manipulations of the mouse inside the laminar flow hood in sterile conditions, it is also acceptable to perform them outside, with a "clean technique", with sterile reagents and surgical instruments in order to avoid contamination.

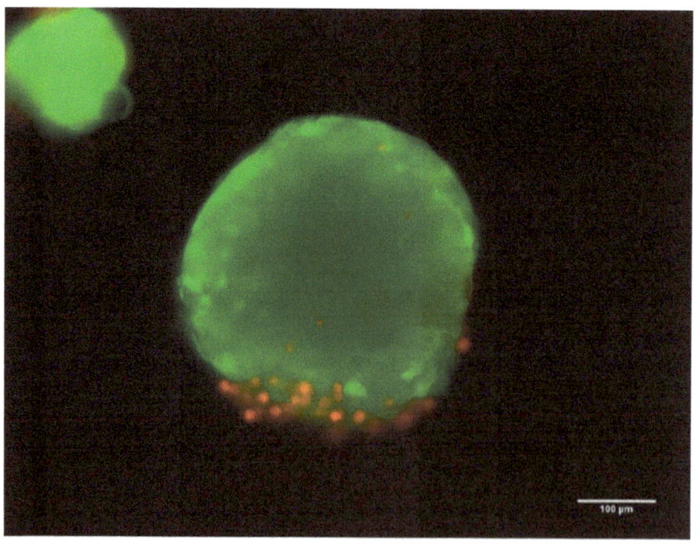

Fig. 3.1 Fluorescence micrograph showing the overall viability of the islets by FDA/PI vital staining. Viable cells are stained in green by FDA and non-viable in red by PI. (*Scale bar* 100 μm)

The mouse is then placed under a stereomicroscope in supine position, with the abdomen cleaned up with 70 % ethanol. A laparotomy is performed, cutting the skin and the muscular tissue of the thorax with a V-incision from the pubic region up to the diaphragm, in order to expose the abdominal cavity. The skin must be well separated from the organs exposed, in order to avoid contamination with the mouse's hair.

The lobes of the liver are then positioned against the diaphragm in order to expose the gall bladder and the proximal segment of the common bile duct. Then, the duodenum is gripped with forceps following the common bile duct and clamped at the level of the Vater ampulla, in order to distend only the pancreas with the collagenase solution. The Vater ampulla is a triangle-shaped white area located at the confluence between the common bile duct and the duodenum.

Once the common bile duct is clamped, it can be cannulated. The standard procedure [7–10] consists in inserting the needle in the Y-shaped junction of the cystic duct and the hepatic duct and injecting the collagenase solution into the common bile duct, distending directly the pancreas. It is important to cannulate well with an optimal needle placement in order to prevent backflow into the liver and drain the splenic tail of the pancreas, an islet rich area. The pancreas is then excised and digested at 37 °C [6, 11]. It is important to remove fat tissue because it may affect digestion and reduce the yield [9].

3.2.2 Enzyme Administration

The key element of this step is the method of the enzyme administration. Originally, the standard procedure set by Lacy et al. [12] was based on the administration of cold saline buffer by the common bile duct to distend the pancreas taking advantage of the anatomical structures, so the enzyme would penetrate and distend better the pancreas, improving the release of the islets from the exocrine tissue. The pancreas was then dissected and the tissue minced in small pieces (1–2 mm), increasing the surface area for the digestion [13]. However, some years later, some modifications to this method were introduced. Gotoh et al. [11] suggested to inject the collagenase in the common bile duct and do a stationary digestion avoiding the mincing step. This is the most common method used nowadays [6–10, 14, 15], since it has a better access with a better digestion of the connective tissue, it produces a yield approximately 50 % higher and it is more cost effective [16].

3.3 Pancreas Digestion

3.3.1 General Considerations About the Procedure

This step entails the digestion of the pancreatic tissue with collagenase once the pancreas is harvested. Usually the tissue is incubated in a water bath at 37 °C, but the duration of the digestion depends on the strength, concentration and formulation of the collagenase. It is also dependent on the strain and age of the mouse [14]. The incubation can be static [11, 12] and/or dynamic [17]; the tissue can be hand-shaken manually (to improve mechanically the dissociation) in the middle of the incubation and/or afterwards, and the tissue can be minced [14, 17] or not [11], depending on the protocol chosen.

In some protocols, the digested tissue is passed through a 14G needle and/or a 450 μm mesh filter, to improve mechanically the dissociation of the tissue. Next, the tissue is washed once or several times before proceeding to the purification phase.

3.3.2 Selection of the Enzyme

The choice of the enzyme is critical. Without a good digestion of the tissue, the purification cannot be effective. Therefore, the knowledge of the collagen composition in the extracellular matrix is crucial for an adequate selection and formulation of the most appropriate enzyme according to the donor's characteristics.

Traditionally, the enzyme used in islet isolation is the bacterial collagenase *Clostridium histolyticum*. The rationale for the use of this enzyme is that collagen is an important component of the pancreatic extracellular matrix (ECM). The use of this collagenase was introduced for the first time by Moskalewski in 1965 [18], and it allows the enzymatic degradation of the ECM and release of islets during the isolation procedure [19].

Traditionally, crude collagenase blends such Collagenase V (Sigma), were used for rodent as well as for human islet isolation [20, 21]. Original crude collagenase preparations from *Clostridium histolyticum* are mixtures of six collagenases, a neutral protease and several enzymes with tryptic-like activity, which also influence the dissociation process [22]. In fact, it has been reported that tryptic-like activity is a key component that optimizes the efficiency of islet isolation, reducing dissociation time and minimizing lot-to-lot variability. The six collagenases are divided in two subtypes: G (or class I) and H (or class II) collagenases. However, there are contradictory studies about their role and importance in human and rodent isolation, probably due to the difference in the extracellular matrix composition between species or the different blends used. Fujio et al. [23] suggested that it is possible that some components of the rat extracellular matrix could only be digested by class II collagenases. Wolters et al. did pioneering work in this regard [24] and reported the predominant role of class II collagenase in rat islet dissociation versus incomplete dissociation obtained with class I collagenase alone. However, Brandhorst et al. reported that the highest yield of rat islet isolation was obtained using the same proportion of class I and class II collagenase (C-ratio of 1.0) [25, 26]. In human islet isolation, collagenase class I is considered essential [27].

With the evolution of this research field, it was observed that crude collagenases, which are derived from bacterial cultures, contained impurities. Key active components often had an imbalanced formulation, there was significant batch-to-batch and vial-to-vial enzyme variability, and high endotoxin levels and pigment contamination were detected [20, 27, 28]. Specifically, endotoxin contamination correlates positively with increased cytokine production and apoptosis and negatively with engraftment in rat islet transplantation models as well as in clinical outcomes [29]. Therefore, the current enzymes used in human islet isolation are purified, despite the suggestion by some authors to use filtrated crude collagenases in human islet isolation to decrease costs [21]. In 2009, Yesil et al. showed a correlation between enzyme purity and yield [30]. The current combinations used for islet isolation consist, mainly, of class I and class II collagenases from *Clostridium histolyticum* and a neutral protease. The neutral protease can be from *Clostridium histolyticum* as well, although the gold standard neutral protease in use is Thermolysin, which is derived from *Bacillus thermoproteolyticus*. The reasons of its success are its low cost, stable production and strong digestion efficacy. However, a recent publication suggests that clostripain (a protease from *Clostridium histolyticum* with tryptic-like activity) could have a synergistic effect with neutral protease and collagenases derived from the same bacteria in rat islet isolation, increasing the efficiency and outcomes of the procedure [31].

In rodent islet isolation, collagenase V (Sigma, Ayrshire, UK), collagenase XI (Sigma, Ayrshire, UK) and collagenase P (Roche, Mannheim, Germany) are routinely used. However, these enzymes are not the only responsible of the tissue dissociation of the pancreas. The pancreas itself is a source of proteolytic endogenous enzymes that are continuously released by the exocrine tissue during the digestion [32]. In fact, Wolters et al. suggested that proteolytic activity caused cell lysis and release of DNA, making the separation of islets from the exocrine tissue difficult. Therefore, they reported that adding inhibitors to the digestion medium to suppress the proteolytic activity, like bovine serum albumin and trypsin inhibitors (purified clostripain, egg white trypsin inhibitor, soybean trypsin inhibitor, etc.) increased the islet yield. However, in a recent study of the effects of some endogenous protease inhibitors (specifically serine protease inhibitors such as Pefabloc, Trasylol and Urinary Trypsin Inhibitor) it was shown that some of them have detrimental effect on the action of bacterial neutral proteases [33]. Pefabloc, in particular (which is widely used in human islet isolation) affects insulin response. In contrast, Urinary Trypsin Inhibitor, which is not yet approved by the FDA, enhances bacterial neutral protease action in addition to inhibiting endogenous proteases. Therefore, the quality and formulation of the digestion enzyme is key for the islet isolation outcomes, both for experimental and clinical islet isolations.

3.3.3 Duration of Digestion

The duration of digestion is dependent, on one hand, on the strength, concentration and formulation of the enzyme; and, on the other, on the strain and age of the animal, according to the guidelines of factors influencing islet isolation published by the Haan [34], since differences in the connective tissue have been observed. Then, the quality of the enzyme and the duration of the digestion constitute a tandem that is critical for the islet isolation outcomes. A prolonged digestion not only causes overdigestion and degradation of the islets but also affects morphology, yield, viability and even functionality [27]. Therefore, due to the batch-to-batch variability in the enzymes used for mouse islet isolation, the duration of the digestion should be carefully optimized prior to any experiment. The typical range of digestion time in mouse pancreas for collagenase P (Roche, Mannheim, Germany) (3 mg/ml) is around 6–7 min and for collagenase V (Sigma, Ayshire, UK) (1 mg/ml) is up to 10 min. This variability is due to the formulation, the concentration, the activity of the enzyme for that batch and the procedure, as well.

3.4 Islet Purification

Once the pancreatic tissue is digested and chemically and mechanically dissociated, the next step is to separate the endocrine of the exocrine tissue and purify the islets. Purification is a key step for islet isolation outcomes as well as for clinical applications, since highly pure preparations lead to engraftment, reduced graft immunogenicity in transplants and suitability for immunomodulation procedures (Fig. 3.2) [6, 7, 35]. Furthermore, an impure preparation can cause important post-surgical complications after intraportal transplantation, such as thrombosis and embolism [36]. While contamination with acinar cells may impair the integrity, viability and functionality of the islets [37], some extra-insular tissue potentially containing progenitor cells could potentially increase islet viability and functionality and improve clinical outcomes as well.

3.4.1 General Considerations About the Procedure

The traditional method used for mouse islet purification is discontinuous density gradient centrifugation (or isopycnic centrifugation) with Ficoll [38]. With this method, the cells in a suspension are separated according to their intrinsic differences in cell density (≈1.059 g/ml for islet tissue and 1.059–1.074 g/ml for exocrine tissue) [39]. The researcher should build a gradient of different layers of Ficoll solutions of different densities, from the densest layer at the bottom of the tube (which is mixed with the cells) to the less dense at the top. The tube is spun at a certain speed and temperature, depending on the density of the cells that we want to purify. Then, each cell type migrates through the gradient to the interface of the layers with the same density. After the centrifugation, the cells are retrieved and washed several times to eliminate the rests of Ficoll.

3.4.2 Density Gradient Purification

The first islet purification was done by Bensley in 1911 [2]. It consisted in hand-picking of islets to separate them manually from the exocrine tissue. However, it was a long and tedious procedure, and not feasible for large-scale experiments. In 1967, Lacy introduced the islet purification by differential density elutriation using the discontinuous sucrose density gradient [12]. However, the osmolarity was very high, due to the very

Fig. 3.2 Purity assessment by dithizone staining. Dark field micrograph of purified islets. (1.25×, *scale bar* 400 µm)

high molar concentration of sucrose in the gradient, and it damaged the islet integrity. Ficoll, a high molecular weight polymer of sucrose (40 kD) was introduced in 1969 by Arnold Lindall [40]. However, since islets obtained by Ficoll purification exhibited impaired insulin secretion, Lacy's team started dialyzing Ficoll with positive results [38].

While bovine serum albumin, percoll or metrizamide have been used in the past [41], the most common discontinuous density gradients used today are Ficoll, Histopaque (a Ficoll-based solution), dextran and iodixanol [42, 43]. Ficoll is the most widely used, as it reduces cell swelling and increases the density differences between islets and exocrine tissue [38]. Although several publications have reported toxic effects of Ficoll on islets, affecting its quality [44–46], more recent studies suggest that neither Ficoll nor Ficoll-based gradients [43, 47] exhibit deleterious effects. In fact, McCall et al. [43] compared different density gradient solutions in terms of islet quality (Ficoll, Histopaque, Dextran and Iodixanol) and showed that the best in terms of islet quality and cost-effectiveness was Histopaque, which is Ficoll-based. Iodixanol is a non-ionic, iso-osmolar solution that was initially used for porcine islet isolation [48] and later applied for rodent [46] and human islet isolation [49]. Iodixanol is currently displacing the use of Ficoll in human islet isolation with the advantage of a lower cost [50, 51].

3.4.3 Other Purification Methods

Although density gradient/isopycnic centrifugation has been the gold standard purification method since the 1960s, other purification methods have also been reported in the literature to purify pancreatic islets: hand-picking [47], phototermolysis [52], radiation [53], differences in osmolality [54], cryo-isolation [55] and cell sorting [56], among others. Specifically, in the case of rodent islet isolation, the main alternative methods that have been reported are: magnetic microspheres coated with islet or cytotoxic anti-acinar monoclonal antibodies [57], osmotic shock [39] and filtration [9, 45, 47]. Ramírez-Domínguez and Castaño [47] reported for the first time a comparison study of Histopaque and filtration purification in terms of quality, time and costs. In the McCall's study [43] Histopaque was recommended over other gradient solutions, but whether islets obtained by filtration had a good quality in comparison with Histopaque-purified islets had not been reported before. According to the above study, islets purified by Histopaque and filtration were of comparable quality, but filtration was more cost-effective because it saved 90 % of the time devoted to purification with Histopaque. Despite anatomical interspecies differences in islets [58–61] the range of islet sizes is similar in mice, humans and other species, with an average diameter of 140 µm and an upper limit of size at around 500–700 µm of diameter [62]. Owing to this, filtration could be considered a method with high translational potential.

3.5 Islet Culture

During the isolation process, islets are affected by many stressing stimuli: "Anoikis" due to the detachment of islets from the surrounding extracellular matrix [63, 64], nutrient and oxygen deprivation, presence of endotoxins, the release of proteolytic enzymes form the acinar tissue, etc. [65]. Therefore, a step of islet culture can be used to restore viability and functionality, improving the islet isolation outcome.

3.5.1 General Considerations About the Procedure

This step could be interpreted as an extension of the purification step, since exocrine tissue does not survive well in culture, enriching therefore the islet fraction (Fig. 3.3). In the culture procedure, aspects such as the choice of culture media and the cell density have to be taken into account.

The most common islet culture medium for murine islets is RPMI 1640 (*Rosewell Park Memorial Institute Medium*). However, some laboratories prefer to use CMRL 1066 (*Connaught Medical Research Laboratories*), which is also widely used to culture human pancreatic islets. Preference for CMRL 1066 is based on the observation that some immune cells, such as dendritic and endothelial cells, cannot grow in that medium, which may therefore induce a decrease in alloreactivity.

One component of the medium that is key for the physiology of the islets is glucose. A concentration of 11 mM glucose resulted in islets with lower apoptosis rates and increased viability than those grown with a different concentration. With glucose concentrations below 11 mM, insulin content was reduced and downregulation of key genes for glucose metabolism was observed. Similarly, glucose concentrations over 11 mM resulted in toxicity [66].

Islet culture media are commonly supplemented with 10 % fetal bovine serum to keep the cell viability and antibiotics/antimycotics to avoid the possibility of contamination.

With reference to other practical issues, murine islets are cultured in suspension dishes,

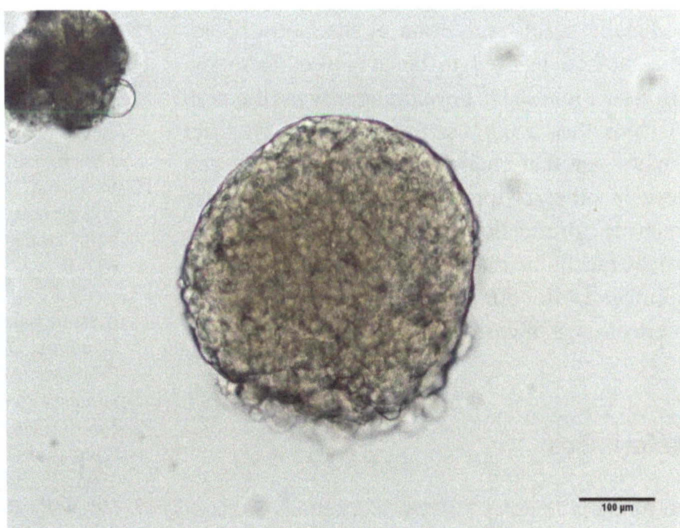

Fig. 3.3 Micograph showing bright field islets. (*Scale bar* 100 µm)

avoiding the growth of acinar tissue and preventing islet attachment. It is also recommended to let islets recover overnight in culture and change the media after 24–48 h post-isolation, to remove debris and prevent cell competition for nutrients. Specifically, it is suggested to seed a maximum of 300 islets in 60 × 15 mm dishes to avoid cell stress and keep optimum culture conditions.

Since the publication of the Edmonton protocol [3], there is some controversy on whether islets should be transplanted fresh or previously cultured. Although some authors have reported the superiority of fresh islets in in vitro and in vivo performance [67], murine islets are generally used after a short culture period.

3.6 Concluding Remarks

This chapter reviews critical aspects of murine islet isolation and highlights the importance of mice as providers of raw material to conduct biological studies on diabetes, with an additional translational potential. This emphasizes the need of implementing an efficient isolation protocol, not only to obtain high quality islets, but also to justify a rational use of animals and laboratory resources. While working with mice is economical and simple, factors such as strain, age, etc. highly influence the isolation outcome. Function, purity and yield are also donor strain-dependent.

Finally, a full translation of the mouse isolation method has yet to be achieved. Research efforts are focused in developing new purification methods that maximize the yield and the islet quality. Another challenge is to investigate and develop other factors discussed in this chapter (enzyme formulation, collagen composition in the extracellular matrix, etc.) with the goal of improving the efficiency and addressing the shortcomings of current protocols.

References

1. American Diabetes Association. Diagnosis and classification of diabetes mellitus. Diabetes Care. 2011;34(Suppl1):S62–9.
2. Bensley R. Studies on the pancreas of the guinea pig. Am J Anat. 1911;12(3):297–388.
3. Shapiro A, Lakey J, Ryan E, Korbutt G, Toth E, Warnock G, et al. Islet transplantation in seven patients with type 1 diabetes mellitus using a glucocorticoid-free immunosuppressive regimen. N Engl J Med. 2000;343(4):230–8.
4. Chakradhar S. Diabetes researchers fear worsening access to human islets. Nat Med. 2014;20(6):567.
5. Kulkarni R, Stewart A. Summary of the keystone islet workshop (April 2014): the increasing demand for human islet availability in diabetes research. Diabetes. 2014;63(12):3979–81.
6. Marzorati S, Ramirez-Dominguez M. Mouse islet isolation. In: Islam S, Editor. Islets of Langerhans. 2nd ed. Dordrecht: Springer; 2015. p. 83–107. ISBN:978-94-007-6685-3.
7. Carter J, Dula S, Corbin K, Wu R, Nunemaker C. A practical guide to rodent islet isolation and assessment. Biol Proc Online. 2009;11(1):3–31.
8. Neuman J, Truchan N, Joseph J, Kimple M. A method for mouse pancreatic islet isolation and intracellular cAMP determination. J Vis Exp. 2014;88:e50374.
9. Li D, Yuan Y, Tu H, Liang Q, Dai L. A protocol for islet isolation from mouse pancreas. Nat Protoc. 2009;4(11):1649–52.
10. Zmuda E, Powell C, Hai T. A method for murine islet isolation and subcapsular kidney transplantation. J Vis Exp. 2011;50:pii: 2096.
11. Gotoh M, Maki T, Kiyozumi T, Satomi S, Monaco A. An improved method for isolation of mouse pancreatic islets. Transplantation. 1985;40(4):437.
12. Lacy P, Kostianovsky M. Method for the isolation of intact islets of Langerhans from the rat pancreas. Diabetes. 1967;16(1):35–9.
13. O'Dowd J. The isolation and purification of rodent pancreatic islets of Langerhans. In: Stocker C, editor. Type 2 diabetes methods and protocols (Methods in Molecular Biology), vol. 560. 1st ed. Totowa: Humana Press; 2009. p. 37–42.
14. Szot G, Koudria P, Bluestone J. Murine pancreatic islet isolation. J Vis Exp. 2007;7:255.
15. Stull N, Breite A, McCarthy R, Tersey S, Mirmira R. Mouse islet of Langerhans isolation using a combination of purified collagenase and neutral protease. J Vis Exp. 2012;67:pii: 4137.
16. Shapiro A. High yield of rodent islets with intraductal collagenase and stationary digestion. A comparison with standard technique. Cell Transplant. 1996;5(6): 631–8.
17. Kelly C, Blair L, Corbett J, Scarim A. Isolation of islets of Langerhans from rodent pancreas. In: Özcan S, editor. Diabetes melitus: methods and protocols (Methods in Molecular Biology). 1st ed. Totowa: Humana Press; 2003. p. 3–15.
18. Moskalewski S. Isolation and culture of the islets of Langerhans of the guinea pig. Gen Comp Endocrinol. 1965;5(3):342–53.
19. McCarthy R, Breite A, Green M, Dwulet F. Tissue dissociation enzymes for isolating human islets for transplantation: factors to consider in setting enzyme

acceptance criteria. Transplantation. 2011;91(2): 137–45.
20. Wolters GH, Vos -Scheperkeuter GH, van Deijnen JH, van Schilfgaarde R. An analysis of the role of collagenase and protease in the enzymatic dissociation of the rat pancreas for islet isolation. Diabetologia. 1992;35(8):735–42.
21. Wang Y, Paushter D, Wang S, Barbaro B, Harvat T, Danielson K, et al. Highly purified versus filtered crude collagenase: comparable human islet isolation outcomes. Cell Transplant. 2011;20(11):1817–25.
22. Brandhorst H, Friberg A, Andersson H, Felldin M, Foss A, Salmela K, et al. The importance of tryptic-like activity in purified enzyme blends for efficient islet isolation. Transplantation. 2009;87(3):370–5.
23. Fujio A, Murayama K, Yamagata Y, Watanabe K, Imura T, Inagaki A, et al. Collagenase H is crucial for isolation of rat pancreatic islets. Cell Transplant. 2014;23(10):1187–98.
24. Wolters GH, Vos-Scheperkeuter GH, Lin H, van Schilfgaarde R. Different roles of class I and class II clostridium histolyticum collagenase in rat pancreatic islet isolation. Diabetes. 1995;44(2):227–33.
25. Brandhorst D, Huettler S, Alt A, Raemsch-Guenther N, Kurfuerst M, Bretzel R, et al. Adjustment of the ratio between collagenase class II and I improves islet isolation outcome. Transplant Proc. 2005;37(8): 3450–1.
26. Brandhorst H, Raemsch-Guenther N, Raemsch C, Friedrich O, Huettler S, Kurfuerst M, et al. The ratio between collagenase class I and class II influences the efficient islet release from the rat pancreas. Transplantation. 2008;85(3):456–61.
27. Barnett M, Zhai X, LeGatt D, Cheng S, Shapiro A, Lakey J. Quantitative assessment of collagenase blends for human islet isolation. Transplantation. 2005;80(6):723–8.
28. van Deijnen J, Hulstaert C, Wolters G, van Schilfgaarde R. Significance of the peri-insular extracellular matrix for islet isolation from the pancreas of rat, dog, pig, and man. Cell Tissue Res. 1992;267(1):139–46.
29. O'Gorman D, Kin T, Pawlick R, Imes S, Senior P, Shapiro A. Clinical islet isolation outcomes with a highly purified neutral protease for pancreas dissociation. Islets. 2013;5(3):111–5.
30. Yesil P, Michel M, Chwalek K, Pedack S, Jany C, Ludwig B, et al. A new collagenase blend increases the number of islets isolated from mouse pancreas. Islets. 2009;1(3):185–90.
31. Dendo M, Maeda H, Yamagata Y, Murayama K, Watanabe K, Imura T, et al. Synergistic effect of neutral protease and clostripain on rat pancreatic islet isolation. Transplantation. 2015;99(7):1349–55.
32. Wolters G, van Suylichem P, van Deijnen J, van Schilfgaarde R. Factors influencing the isolation process of islets of Langerhans. Horm Metab Res Suppl. 1990;25:20–6.
33. Nduaguibe C, Bentsi-Barnes K, Mullen Y, Kandeel F, Al-Abdullah I. Serine protease inhibitors suppress pancreatic endogenous proteases and modulate bacterial neutral proteases. Islets. 2010;2(3):200–6.
34. de Haan B, Faas M, Spijker H, van Willigen J, de Haan A, de Vos P. Factors influencing isolation of functional pancreatic rat islets. Pancreas. 2004;29(1): e15–22.
35. Lakey J, Burridge P, Shapiro A. Technical aspects of islet preparation and transplantation. Transpl Int. 2003;16(9):613–32.
36. Gray D, Sutton R, McShane P, Peters M, Morris P. Exocrine contamination impairs implantation of pancreatic islets transplanted beneath the kidney capsule. J Surg Res. 1988;45(5):432–42.
37. Loganathan G, Dawra RK, Pugazhenthhi S, Guo Z, Soltani SM, Wiseman A, Sanders MA, Papas KK, Velayutham K, Saluja AK, Sutherland DE, Hering BJ, Balamurugan AN. Insulin degradation by acinar cell proteases creates a dysfunctional environment for human islets before/after transplantation: benefits for α-1 antytripsin treatment. Transplantation. 2011;92(11):1222–30.
38. Scharp D, Kemp C, Knight M, Ballinger W, Lacy P. The use of Ficoll in the preparation of viable islets of Langerhans from the rat pancreas. Transplantation. 1973;16(6):686–8.
39. Li N, Zhang Y, Liu Y, Xiu Z, Tan M, Li S, et al. Improved islet purity by the hypertonic-hypotonic method. IJAO. 2014;37(6):477–85.
40. Lindall A, Steffes M, Sorenson R. Immunoassayable insulin content of subcellular fractions of rat islets 1. Endocrinology. 1969;85(2):218–23.
41. Suylichem P, Wolters G, Schilfgaarde R. The efficacy of density gradients for islet purification: a comparison of seven density gradients. Transpl Int. 1990;3(1):156–61.
42. Kaufman D, Sutherland D. Pancreatic islet isolation in rats with ductal collagenase distention, stationary digestion and dextran separation. Transplantation. 1988;45(2):493–4.
43. McCall M, Maciver A, Pawlick R, Edgar R, Shapiro A. Histopaque provides optimal mouse islet purification kinetics: comparison study with Ficoll, iodixanol and dextran. Islets. 2011;3(4):144–9.
44. Lake S, Anderson J, Chamberlain J, Gardner S, Bell P, James R. Bovine serum albumin density gradient isolation of rat pancreatic islets. Transplantation. 1987;43(6):805–8.
45. Salvalaggio P, Deng S, Ariyan C, Millet I, Zawalich W, Basadonna G, et al. Islet filtration: a simple and rapid new purification procedure that avoids Ficoll and improves islet mass and function. Transplantation. 2002;74(6):877–9.
46. Delle, Saito M, Yoshimoto P, Noronha I. The use of iodixanol for the purification of rat pancreatic islets. Transplant Proc. 2007;39(2):467–9.

47. Ramirez-Dominguez M, Castaño L. Filtration is a time-efficient option to Histopaque, providing good-quality islets in mouse islet isolation. Cytotechnology. 2015;67(2):199–206.
48. van der Burg M, Basir I, Bouwman E. No porcine islet loss during density gradient purification in a novel iodixanol in university of Wisconsin solution. Transplant Proc. 1998;30(2):362–3.
49. Noguchi H, Ikemoto T, Naziruddin B, Jackson A, Shimoda M, Fujita Y, et al. Iodixanol-controlled density gradient during islet purification improves recovery rate in human islet isolation. Transplantation. 2009;87(11):1629–35.
50. Mita A, Ricordi C, Miki A, Barker S, Khan A, Alvarez A, et al. Purification method using iodixanol (OptiPrep)-based density gradient significantly reduces cytokine chemokine production from human islet preparations, leading to prolonged cell survival during pretransplantation culture. Transplant Proc. 2009;41(1):314–5.
51. Mita A, Ricordi C, Messinger S, Miki A, Misawa R, Barker S, et al. Antiproinflammatory effects of iodixanol (OptiPrep)-based density gradient purification on human islet preparations. Cell Transplant. 2010;19(12):1537–46.
52. Brunicardi F, Suh E, Kleinman R. Selective photodynamic laser treatment of dispersed pancreatic tissue for islet isolation. Transplant Proc. 1992;24(2):2796.
53. Nason R, Rajotte V, Procyshyn A. Purification of pancreatic islet cell grafts with radiation. Transplant Proc. 1986;18:174.
54. Shimoda M, Itoh T, Sugimoto K, Takita M, Chujo D, Iwahashi S, et al. An effective method to release human islets from surrounding acinar cells with agitation in high osmolality solution. Transplant Proc. 2011;43(9):3161–6.
55. Taylor M, Baicu S. 91. Cryo-isolation: a novel new method for enzyme-free isolation of pancreatic islets involving in situ cryopreservation of islets and selective destruction of acinar tissue. Cryobiology. 2011;63(3):331.
56. Jiao L, Gray D, Gohde W, Flynn G, Morris P. In vitro staining of islets of Langerhans for fluorescence-activated cell sorting. Transplantation. 1991;52(3): 450–2.
57. Fujioka T, Terasaki P, Heintz R, Merideth N, Lanza R, Zheng T, et al. Rapid purification of islets using magnetic microspheres coated with anti-acinar cell monoclonal antibodies. Transplantation. 1990;49(2): 404–7.
58. Brissova M, Fowler MJ, Nicholson WE, Chu A, Hirshberg B, Harlan DM, Powers AC. Assessment of human pancreatic islet architecture and composition by laser scanning confocal microscopy. J Cyotchem Histochem. 2005;53(9):1087–97.
59. Cabrera O, Berman DM, Kenyon NS, Ricordi C, Berggren PO, Caicedo A. The unique cytoarchitecture of human pancreatic islets has implications for islet cell function. PNAS. 2006;103(7):2334–9.
60. Bonner-Weir S, Sullivan BA, Weir GC. Human islet morphology revisited: human and rodent islets are not so different after all. J Histochem Cytochem. 2015;63(8):604–12.
61. Dolenek J, Rupnik M, Stoer A. Structural similarities and differences between the human and the mouse pancreas. Islets. 2015;7(1):e1024405.
62. In't Veld P, Smeets S. Microscopic anatomy of the human islet of Langerhans. In: Islam S, editor. Islets of Langerhans. 2nd ed. Dordrecht: Springer; 2015. p. 19–38.
63. Wang R. Maintenance of beta-cell function and survival following islet isolation requires re-establishment of the islet-matrix relationship. J Endocrinol. 1999;163(2):181–90.
64. Ilieva A. Pancreatic islet cell survival following islet isolation: the role of cellular interactions in the pancreas. J Endocrinol. 1999;161(3):357–64.
65. Marzorati S, Antonioli B, Nano R, Maffi P, Piemonti L, Giliola C, et al. Culture medium modulates proinflammatory conditions of human pancreatic islets before transplantation. Am J Transplant. 2006;6(11):2791–5.
66. Robertson R, Harmon J, Tran P, Tanaka Y, Takahashi H. Glucose toxicity in Â -cells: type 2 diabetes, good radicals gone bad, and the glutathione connection. Diabetes. 2003;52(3):581–7.
67. Takahashi H, Goto M, Ogawa N, Saito Y, Fujimori K, Kurokawa Y, et al. Superiority of fresh islets compared with cultured islets. Transplant Proc. 2009;41(1):350–1.

Pancreatic Islets: Methods for Isolation and Purification of Juvenile and Adult Pig Islets

4

Heide Brandhorst, Paul R.V. Johnson, and Daniel Brandhorst

Abstract

The current situation of organ transplantation is mainly determined by the disbalance between the number of available organs and the number of patients on the waiting list. This obvious dilemma might be solved by the transplantation of porcine organs into human patients. The metabolic similarities which exist between both species made pancreatic islets of Langerhans to that donor tissue which will be most likely transplanted in human recipients. Nevertheless, the successful isolation of significant yields of viable porcine islets is extremely difficult and requires extensive experiences in the field. This review is focussing on the technical challenges, pitfalls and particularities that are associated with the isolation of islets from juvenile and adult pigs considering donor variables that can affect porcine islet isolation outcome.

Keywords

Pig islet isolation • Pig islet purification • Porcine donor variables • Islet xenotransplantation

H. Brandhorst (✉) • D. Brandhorst
Nuffield Department of Surgical Sciences, University of Oxford, Churchill Hospital, Old Road, Oxford OX3 7LJ, UK

Oxford Centre for Diabetes, Endocrinology and Metabolism, Oxford, UK
e-mail: Heide.Brandhorst@nds.ox.ac.uk

P.R.V. Johnson
Nuffield Department of Surgical Sciences, University of Oxford, Churchill Hospital, Old Road, Oxford OX3 7LJ, UK

Oxford Centre for Diabetes, Endocrinology and Metabolism, Oxford, UK

Oxford NIHR Biomedical Research Centre, Oxford, UK

M. Ramírez-Domínguez (ed.), *Pancreatic Islet Isolation*, Advances in Experimental Medicine and Biology 938, DOI 10.1007/978-3-319-39824-2_4

4.1 Introduction

While the early era of modern organ transplantation was challenged by the rapid rejection of the transplanted tissue [1], one of the main problems of organ transplantation in the current year is the imbalance between the number of available donor organs and the number of patients on the waiting list. Recent data from the North American United Network for Organ Sharing (UNOS) show that the donation rate increased by 94 % over a period of more than 20 years. This significant increase has been nullified by the simultaneous expansion of the waiting list by 253 %. The worldwide equivalent figure, as provided by the World Health Organization, is an organ supply-to-demand ratio of 1:25 [2].

This obvious dilemma could potentially be solved by the transplantation of xenogeneic organs into human patients. Due to the anatomical and physiological similarities between humans and pigs, porcine donors are the species of choice to fill the gap between the demand and supply of human organs. However, the metabolic and endocrine concordance between both species made pancreatic islets of Langerhans to that donor tissue which will be most likely transplanted in human recipients in clinical trials [2–4]. In fact, the first series of clinical islet xenotransplantation was performed in 1993 and involved transplanting fetal pig islets into ten insulin-dependent patients with type 1 diabetes. Although small amounts of porcine C-peptide could be detected in the urine for more than 1 year, a reduction of exogenous insulin demand was not achieved in any of the recipients [5]. Subsequent approaches provided much more promising results demonstrating viable tissue nearly 10 years after transplantation of encapsulated neonatal pig islets into a non-immunosuppressed patient with type 1 diabetes [6]. The finding that the vigorous antibody- and complement-mediated rejection that is observed after pig-to-primate xenotransplantation [1, 7] can be reduced using adequate immunosuppression [8, 9], anti-clotting treatment [10] or multi-transgenic donor pigs [11–13] makes clinical trials for pig islet xenotransplantation a realistic option for providing a widespread treatment for a larger population of patients with type 1 diabetes [14].

However, before treatment of type 1 diabetic patients with pig islets becomes a daily reality, numerous precautions must be taken to establish islet xenotransplantation as a safe procedure. Specific pathogen-free (SPF) breeding in biosecure facilities is a mandatory measure to prevent transmission of zoonotic pathogens to the human population [15]. The costs of running these highly specialised facilities will have an impact on the total costs for each islet isolation that are likely to exceed the costs for human islet allotransplantation [16], particularly if pancreases from adult pigs older than 2 years are processed.

The immunological, technical and economical advantages and disadvantages of using pancreases retrieved from fetal, neonatal, juvenile or adult pigs have been extensively discussed [17–19]. This review will rather focus on the technical challenges, pitfalls and specifics that are associated with the isolation of islets from the pancreas of pigs of approximately 6 months to older than 2 years.

4.2 Pig Donor Variables

4.2.1 Pig Strain

Islet isolation always starts with donor selection. The selection of pig strains suitable for successful release of intact islets from the pancreas is of relevance for islet research but is of outstanding importance for clinical xenotransplantation considering the immense costs related to breeding and maintenance of SPF pigs in biosecure facilities.

The very few histomorphologic studies in native pig pancreases which have been performed so far clearly revealed that significant differences in terms of islet size, shape and number exist between different pig strains such as German Landrace, Duroc, Piétrain or wild boars. In two studies it was found that pancreases of wild boars have a high density of islets of very small diameter while the German Landrace is characterized by a high percentage of large islets resulting in low and high total islet volume, respectively [20, 21]. Duroc and Piétrain pigs seem to have an islet volume in between these extreme poles.

However, to categorize the suitability of different strains for successful islet isolation the expression of the peri-islet connective tissue matrix has to be considered as well. Two studies reported in agreement that Duroc pigs have a low expression of the peri-islet capsule while the German Landrace and the German Large White Pig (Deutsches Edelschwein) are characterized by a strong expression of this structure [20, 22] that seems to protect islets from the harmful effects of proteolytic enzymes used for pancreas digestion. Despite the differences that exist in the thickness and perimeter of the peri-islet capsule it is remarkable how similar the capsular composition of different extracellular matrix (ECM) proteins is in widely disparate strains such as Piétrain, Goettinger Minipig or wild boar [22].

We are aware of only a few studies that have correlated the strain of the donor pigs with porcine islet isolation outcome. In agreement with the histomorphologic studies of Kirchhof et al. and Meyer et al. [20, 22] we found a clear advantage of isolating islets from the pancreas of adult Large White Pigs (Deutsches Edelschwein) compared to German Landrace and hybrid pigs [23]. From two other reports we can conclude that Piétrain pigs provided higher islet yields compared to German Landrace and hybrid pigs [21, 24].

There is no doubt that the selection of the pig strain has a significant effect on islet isolation outcome. However, to rate the relevance of these previous observations for the current situation it is important to realize that pig breeding is a continuously ongoing process performed to meet the demands of the consumers [25]. For French Large White Pigs it was demonstrated for the period between 1977 and 1998 that the lean muscle mass increased by 15 % while the relative liver weight decreased by 16 % [26]. Since no data are available for the pancreas weight we can only speculate that the changes in relative fat and muscle mass may also affect the pancreas as well as islet cell growth and proliferation. However, it has been clearly shown in different livestock species that a high lean muscle mass inversely correlates with the fat content of the carcass which is mainly under the control of the insulin-glucose metabolism [27]. In agreement, pigs with a lean phenotype have a relatively higher somatotropin serum level compared to pigs with an obese phenotype which are characterized by a relatively higher insulin level or a larger insulin response toward stimuli [28–31]. Regardless of ease of islet separation, it can be anticipated that the islet volume within the pancreas of pigs is continuously reduced as long as pig breeding is mainly aiming on a higher percentage of lean muscle mass in the carcass.

Within a short period of approximately 3 years (2006–2008) we observed a significant increase of the variability in isolation outcome when islets were isolated from the pancreas of BHZP (German Federal Hybrid Breeding Programme) pigs. As shown in Fig. 4.1, this observation correlated with an alteration in the morphology of islets assessed in the pancreas of donor pigs of the identical strain. While samples retrieved before 2006 showed native islets with an ovoid shape well demarcated from acinar tissue (Fig. 4.1a), islets assessed in the same strain after 2008 were characterized by a smaller diameter or a dumbbell-like shape (Fig. 4.1b).

However, when considering the enormous costs for housing, breeding and maintenance of adult SPF strains [15, 16], miniature pigs will become more and more important as potential donors for clinical islet xenotransplantation. Future efforts should therefore be aimed to crossbreed obese farm pig strains identified as effective islet donors withsuitable strains of miniature pigs [32].

4.2.2 Porcine Donor Age

One important variable to be discussed in the context of donor selection is the donor age. First attempts to isolate islets from the pancreas of market age pigs revealed an unique fragility of islets which appears to be more pronounced than in any other species investigated so far. During

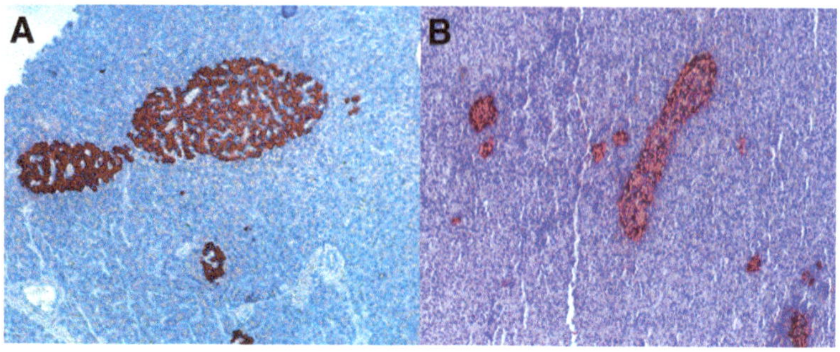

Fig. 4.1 Insulin immunostaining of native pancreas sections retrieved from BHZP (German Federal Hybrid Breeding Programme) retired breeder pigs before 2006 (**a**) and after 2008 (**b**) (magnification × 10)

the isolation procedure juvenile islets fragmented into small clusters or single cells [33]. This may be explained by the weak expression of ECM proteins in pig pancreases, particularly in the peri-islet region, when compared to other species such as rat, dog and human [34, 35]. The fragmentation of pig islets may be accelerated due to the particular specific cytoarchitecture of pig islets composed of several smaller subunits with a beta-cell core and an alpha-cell mantle [36]. Further studies have indicated that islets isolated from adult pigs are morphologically more stable and are easier isolated compared with juvenile pig islets [32, 37, 38].

Although individual differences exist in young and old pigs [39, 40] pancreases from retired breeders contain a higher frequency of larger and compact islets compared with juvenile pig pancreases which are defined by a significantly higher density of islets partially appearing as small clusters composed of only a few cells [21, 41–43]. The aging process does not only contribute to increased islet size but seems also to enhance the total collagen content in pig pancreases which may reflect the age-dependent formation of a protective matrix of ECM proteins around islets [44]. Indeed, we (unpublished data) and others observed the presence of a peri-islet capsule in adult pig pancreases but not in young pigs [38]. In agreement, in a small number of samples it was found that the expression level of laminin, fibronectin, collagen type I, III and IV is higher in old pigs compared to young pigs [22]. In contrast, another report revealed that the pancreas of juvenile pigs contains a higher total collagen amount than adult pigs [45]. Subsequent studies could not confirm a difference between young and old pigs regarding the formation of a peri-islet capsule or the histologic distribution of different collagen types. Among all collagen types assessed, collagen VI was the most abundant intra- and peri-islet collagen type [46]. This is particularly remarkable as collagen VI seems to be resistant towards cleavage by bacterial crude collagenase [47]. However, another study could not detect any collagen VI in or around islets collected from different strains [22].

Nevertheless, it seems to be important to note that the age of the donor pig does not only determine the separability of islets but also the in vitro as well as in vivo function of the islets. It was shown that a higher islet volume of juvenile pig islets is needed to restore normoglycemia in diabetic rodents when compared with posttransplant function of adult pig islets [37, 41, 42, 48]. This, again, seems to reflect the morphological disintegration of juvenile pig islets at the periphery caused by the isolation process and demonstrated by the loss of peripheral alpha cells [49, 50].

Overall, the majority of studies clearly demonstrate that adult pig pancreases are characterized by a larger number of compact islets and a larger functional islet capacity compared to juvenile donors, but the data about the expression of ECM proteins are very inconsistent. The contradiction between different reports may reflect the use of heterogeneous tissue samples and the application of non-specific ineffective antibodies.

4.3 Pig Pancreas Procurement

The method of choice for pig pancreas procurement depends on the scientific purpose and on the financial resources available. Generally, two different strategies are available to procure a pig pancreas. While the slaughterhouse procedure is straightforward, it carries the risk of extensive bacterial contaminations and exposure to significant warm ischaemia. However, a surgical approach is complex and associated with enormous costs. The logistical requirements of a pig pancreas procurement procedure according to clinical standards seem to favour the selection of miniature pigs as potential donors for islet xenotransplantation in patients with type 1 diabetes.

Nevertheless, apart from financial and logistical considerations the few reports published so far indicate the detrimental effects of a surgical pancreas procurement on pig islet isolation outcome [51, 52] when compared with a slaughterhouse procedure that involves immediate brain death induction, exsanguination and rapid evisceration to minimize warm ischaemia time [53]. Whether these observations are related to the use of unsuitable perfusion media, excessive perfusion pressure or to physiological stress caused by extensive medication and surgical trauma remains unclear [54].

Regardless which technique is selected for pig pancreas procurement, our own experience, as well as that of others, clearly indicates that the removal of blood from pancreatic tissue is absolutely essential to ensure proper activity of intraductally perfused enzymes. The underdigestion of pancreatic tissue, resulting from the inhibitory effect of residual blood on enzyme activity, can cause a substantial reduction in islet yield [55, 56].

4.3.1 Intraductal Pancreas Perfusion Technique

Intraductal pancreas perfusion is a backbench procedure that has a duplicate function: first, it decreases rapidly the core temperature of the pancreas in order to reduce detrimental effects of warm ischaemia without the need for intravascular flushing [57]. Second, it is the most efficient way to introduce the enzyme blend into the pancreas.

Manual distension of the pancreas using a syringe is easy to apply as it does not require special equipment. In contrast, the automated pancreas perfusion is performed by means of a complex pump-perfusion device offering the option of continuous pressure control [58, 59]. However, the choice of the distension method depends also on the timing of intraductal pancreas perfusion. Early intraductal pancreas perfusion performed immediately after pancreas retrieval in a slaughterhouse or in a surgical animal facility prior to pancreas shipment to the isolation facility is mostly a logistical challenge that is only be feasible by using a minimum of equipment.

Comparison of manual distension with automated perfusion in human pancreases, has demonstrated the superiority of controlled perfusion over manual syringe loading with regard to islet yield [60]. Although similar comparative experiments have not been performed in the pig pancreas, the findings of Lakey et al. may be particularly relevant for the soft and fragile porcine pancreatic tissue. At 180 mmHg of perfusion pressure the perfusion medium penetrates islets and expandes the intercellullar space thereby loosening islet structure and rendering pig islets more sensitive toward disintegration [61]. The penetration of pig islets can only be prevented by using an intraductal perfusion pressure of less than 50 mmHg [62].

4.3.2 Intraductal Pancreas Perfusion Buffer

The selection of the most suitable pancreas perfusion medium and/or collagenase solvent is primarily determined by the duration of cold ischaemia time until pig pancreas digestion can be initiated. The current standard medium that is mostly used as enzyme buffer is Hank's balanced salt solution (HBSS). However, the simple composition of HBSS does not allow the preservation of tissue during prolonged cold ischaemia. This is particularly relevant for cold stored pig pancreases which have a lower tolerance for ischemically induced damage than human pancreases [63, 64]. This specific sensitivity for the detrimental effects of ischaemia does not only concern the integrity of acinar tissue [65] but also of the ductal system. The low content of fibrous tissue in the pig pancreas accelerates the collapse of the ductal system during cold storage which prevents efficient intrapancreatic enzyme distribution at a later time point [66]. Measurements in adult pig pancreases demonstrated that the integrity of the ductal system is also affected by warm ischaemia. It was noted that the intraductal pressure was more than doubled in perfused pig pancreases when exposed to warm ischaemia longer than 20 min [56]. These findings correspond to improved homogenous distribution of infused collagenase and increased islet yield in human pancreases when the ductal system was initially stabilized by intraductal perfusion of Kyoto preservation solution prior to cold storage [67]. Similar observations were made in pig pancreases using University of Wisconsin (UW) solution for intraductal perfusion [57]. These findings clearly suggest that the early intraductal perfusion/distension of the pancreas with UW-solution is particularly relevant when islets are isolated from pig pancreases exposed to significant cold ischaemia [68].

Studies in adult pig pancreases revealed that the replacement of HBSS by UW-solution administered as intraductal enzyme buffer increased islet yield or islet viability even in pancreases processed after short cold ischaemia time [37, 69–71]. In perfect agreement with Heald et al. [70] we found that collagenase concentration has to be increased by 25 % to optimize release of cleaved islets when using UW-solution [37] which is in accordance with the observation that UW-solution inhibits collagenase class II activity [72]. Because no data are currently available, we can only speculate whether UW-solution may also affect collagenase class I activity. As a consequence of these considerations different modified or simplified versions of UW-solution were utilized for intraductal pig pancreas collagenase perfusion [71, 73] aiming on omission of potentially inhibitory components such as high potassium, magnesium, adenosine, allopurinol, glutathione and hydroxyethylstarch [74].

The decision whether the collagenase blend has to be administered prior to or after cold storage of pig pancreases seems to depend on the preservation procedure used. While intraductal enzyme perfusion prior to static cold storage improves islet isolation outcome from ischaemic pig pancreases [68, 75], the intraductal administration of enzymes prior to prolonged pig pancreas preservation by means of the two-layer method seems to be detrimental for porcine islet integrity and viability [64].

4.4 Pig Pancreas Digestion

The primary goal of pancreas digestion is the release of a maximum number of intact islets from within the surrounding acinar tissue. The requirement to obtain complete dissociation of the non-endocrine components of the pancreas without dispersing the endocrine cell clusters is unique in the field of tissue separation [76] and reflects the dilemma of any effective islet isolation method to balance two opposite priorities. To achieve this delicate equilibration the vast majority of procedures combine the application of collagenolytic enzymes with mechanical treatment of the gland.

4.4.1 Enzyme Blend Selection

Since the structure and consistency of the acinar tissue differs enormously between human and porcine pancreases, enzyme activities need to be adjusted for the digestion of the porcine pancreas which appears to be much more fragile in comparison with the human pancreas. In previous studies we found that the concentration of a crude collagenase mixture needed to efficiently digest a human pancreas has to be reduced by nearly 50 % for a retired breeder pig pancreas. Moreover, the enzyme activities were additionally reduced by decreasing the digestion temperature from 37 to 32 °C [37]. Nearly identical findings were observed when using purified enzyme blends for adult pig pancreas digestion. While the enzyme concentration was reduced in a range from 50 to 67 % compared to the human concentration, the digestion temperature was adjusted between 24 and 28 °C [77] which is equivalent to a further reduction of the enzyme activity by 50 % [78].

However, the strategy to reduce enzyme activity by decreasing the digestion temperature also has the advantage of reducing the metabolic activity and nutritive demand of islets that are exposed to a detrimental environment. This environment is characterized by anoxia [79], acid pH, hyperosmolarity and the presence of harmful endogenous enzymes [80–82] thereby reducing the morphological and functional integrity of islets. Other attempts have aimed to inhibit trypsin as initial activator of the endogenous zymogen cascade. Although large variability in trypsin activation seems to exist between individual pig pancreases [82], the majority of studies indicate an improvement of islet yield or post purification recovery when the intraductally administered enzyme perfusion medium was supplemented with trypsin inhibitors [83–86].

Previous studies have clearly demonstrated that neutral protease from Clostridium histolyticum or Bacillus thermoproteolyticus rokko is detrimental for islet integrity [87–89] particularly when used for ischaemic pig pancreases [64]. To reduce the harmful effects of proteolytic enzyme blends, neutral protease was replaced by dispase in the Liberase PI blend (Roche Diagnostics, Indianapolis, USA) specifically developed for porcine islet isolation [90]. Dispase has been shown to be less harmful during dispersion of isolated rat islets when compared with other enzymes [91]. This observation may have implications for the isolation of intact islets as well. As dispase does not cleave laminin (a vital component of the basement membrane [92, 93]), it can be hypothesized that the loss of the basement membrane, that is usually observed during enzymatic islet isolation [94], is reduced thereby improving islet survival during the isolation procedure, after culture and posttransplant. This enzyme blend has been established as widely used enzyme blend for pig islet isolation [39, 56, 73].

Nevertheless, because evidence is still lacking that dispase has significant advantages compared with neutral protease, our current protocol for adult pig islet isolation utilizes a two-component enzyme blend composed of collagenase and neutral protease (Serva Electrophoresis, Uetersen, Germany) that offers the option to individually reduce the neutral protease activity according to donor variables, pancreas procurement and cold ischaemia time [95, 96].

4.4.2 Dissociation

In the history of islet isolation, several different techniques have been established and then discontinued as soon as a more advanced method was developed [97]. During this evolutionary process, the automated digestion-filtration method developed by Ricordi still representing the current gold standard for the dissociation of human pancreases for islet release [98]. The automated digestion-filtration device consists of a continuously shaken digestion chamber loaded with the enzyme-distended pancreas together with a number of stainless-steel balls to increase the mechanical dissociation of the tissue. The chamber is topped by a 500 µm-mesh and closed by a conical top. This digestion-filtration device is implemented in a closed circuit during the recirculation phase to allow continuous perfusion of the system with different buffers from the flat bottom of the chamber to its conical top. As soon as fluid samples taken from the digestion circuit demonstrate release of a significant number of cleaved islets the collection phase of digestion is started to recover free islets from enzyme activity. While released islets are collected, the mechanical treatment of the pancreas in the agitating chamber is ongoing as non-digested tissue is retained by the mesh. This system has the great advantage that a final endpoint of the digestion does not need to be determined. This is in contrast to static digestion procedures as practised in the past [99].

Although this technique has been established in all centres for clinical islet transplantation, numerous variations have been introduced over time. These modifications mainly concern the replacement of motor-driven movement by manual agitation of the digestion chamber, the amplitude and frequency of shaking as well as the spatial orientation of top and bottom chamber side during agitation. Our own experiences in porcine islet isolation resulted in a mixture of automated chamber shaking at the beginning of the recirculation phase and manual agitation at the end of recirculation and during collection phase. During the manual agitation the chamber is frequently turned upside down to facilitate and optimize the release of cleaved islets by means of gravity force which appears to be important because the volume of a distended adult pig pancreas can entirely fill the digestion chamber [77, 95, 96].

However, the variability of techniques for processing pig pancreases is much broader than what has been described for human pancreases and reflects the difficulties successfully releasing intact islets from pig pancreases. The selection of the dissociation technique seems to depend mostly on the age and consistency of the donor tissue (see previous section). While the automated digestion-filtration method seems to provide excellent isolation outcomes when pancreases from suitable retired breeder pigs are dissociated [56, 77, 100] its mechanical action seems to be too harsh for pancreases from market age pigs.

Therefore, a non-traumatic static digestion procedure was established for young donor pigs [73] resembling on early attempts to isolate islets from human pancreases [99]. These methods use a closed or open container for pancreas incubation at 37 °C. Mechanical dissociation is performed at later stages of the enzymatic digestion using forceps and scissors to tear predigested tissue apart. Because these digestion procedures are characterized by a fixed endpoint they are bearing the significant risk for under- as well as overdigestion. Moreover, as discussed in the previous section, the prolonged exposure of islets to anoxia [79], hyperosmolarity and harmful endogenous enzymes [80–82] during digestion can significantly reduce viability and function of islets. The early recovery of released islets from this environment seems to be absolutely essential to preserve morphological and functional integrity of islets [101]. An elegant prospective study, comparing the digestion of juvenile pig pancreases that were longitudinally splitted into two identical portions of the splenic lobe, clearly demonstrated the superiority of the automated digestion-filtration method when compared to a static manual digestion procedure [102].

Inspired by an early version of a digestion-filtration chamber, designed by Scharp et al. for digestion of the pancreas from rhesus monkeys [103], we developed a preliminary model of an oxygen-flushed chamber in order to minimize the traumatic shear stress without completely omitting mechanical treatment of the sensitive pig pancreas. As shown in Fig. 4.2 the enzyme-distended pancreas is loaded into a basket made from a 500 μm-pore-sized mesh fixed in the center of a continuously oxygen-flushed chamber. To obtain a significant agitation of the pancreas an oxygen flow of 300–500 mL/min is used. Overpressure is released through a vent in the top of the chamber. During the recirculation and collection phase the chamber is continuously perfused in a reversed direction from its top to its conical bottom supporting sampling and collection of released islets by gravity. Initial pilot studies indicated an effective release of islets with fully preserved morphology (unpublished data).

4.5 Porcine Islet Purification

Separation of isolated islets from exocrine cells, lymph nodes, vascular and ductal components of the pancreas is essential prerequisite for subsequent transplantation of islets into patients with type 1 diabetes. Islet purification is essentially needed to decrease the tissue volume which is important to limit the portal pressure [104–106] and the risk of portal vein thrombosis [107, 108] after intraportal islet infusion in the recipient. Moreover, the elimination of non-islet tissue

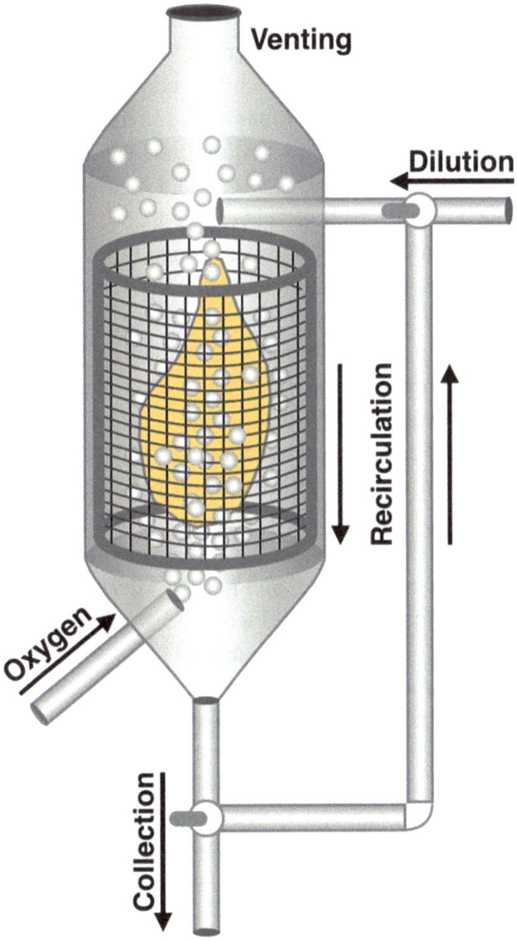

Fig. 4.2 Continuous digestion-oxygenation device

reduces the immunogenicity of the islet preparation [109, 110] and enables successful islet culture [111–113] which is required to perform islet quality assessment and patient management prior to transplantation [114, 115].

In several animal models it has been clearly shown that exocrine tissue induces necrosis and impairs engraftment of syngeneic or autologous islets when transplanted under the kidney capsule, into the liver or into the intramuscular site [116–118]. These observations were confirmed after intraportal autotransplantation of human islets [119] and may be even more relevant for islet xenotransplantation where implanted tissue provokes a stronger response than after autologous and allogeneic islet transplantation [120–122].

The principle of separation of different cell populations is based on the tissue-specific density and on the diameter of isolated cellsor tissue fragments [123]. While it might be possible to control the size of digested pancreatic tissue in highly standardized donors such as inbred mice [124], in porcine and human donors the diameter of islets and the size of digested exocrine tissue fragments vary enormously from one preparation to the other but also within the same preparation. Research on islet purification techniques has therefore been mainly focused on media and techniques that are most efficient to separate islets from non-islet tissue according to the tissue-specific density.

4.5.1 Determinants of Intrinsic Pig Pancreas Density

It is obvious, that isopycnic separation of heterogeneous cell suspensions to obtain homogeneous cell populations can only successfully work when a difference exists between the density of exocrine particles and isolated islets. In fact, the specific density of acinar and islet cells differs to a small extent [125] but can be altered in the pig by several factors. One of the most important variables that influence the specific density of the exocrine part of the pancreas is the nutritional status of the donor pig. During one hour of stimulation, islets can secrete a maximum of only 10 % of the initially stored insulin [126], while acinar cells can discharge more than 40 % of the exocrine zymogen granules [127–129]. Because of their high molecular weight any change in the intracellular protein level has a strong impact on the density of exocrine tissue [123, 130]. Vice versa, during fasting acinar cells discharge only a marginal amount of enzymes and have a higher density compared to postprandial conditions [131]. These findings may justify the advice to retrieve pancreases only from donor pigs after overnight fasting prior to pancreas procurement.

On the other hand, any incident that causes oedematous cell swelling simultaneously reduces the density of exocrine tissue particularly after prolonged cold ischaemia time, after pancreas perfusion with large volumes of organ preservation media and subsequent to enzymatic pancreas digestion [132, 133]. The significant overlap of the specific densities of exocrine and islet tissue deteriorates the successful purification of isolated islets.

4.5.2 Reversing Cell Swelling

Oedematous cell swelling is a phenomenon that occurs particularly at hypothermia when the sodium-potassium pumps are arrested and the membrane permeability for sodium and potassium is increased [134]. This phenomenon may vary between different tissues but can be observed in all cold-stored organs [135]. Moreover, cell swelling may continue or even increase after rewarming of the tissue [136].

It was clearly demonstrated in canine and later in human pancreases that cold storage-related cell swelling can be reversed by washing and incubation of the pancreatic digest in UW-solution thereby maintaining or enhancing the difference in the specific density between exocrine tissue and islets [137–139]. The principle of incubating digested pancreatic tissue in UW-solution prior to density gradient centrifugation was confirmed and established for purification of isolated pig islets [37, 140]. As a further development, van der Burg et al. utilized UW-solution as medium for all steps of pig pancreas processing including collagenase perfusion, pancreas digestion, and purification [69].

Due to the complexity of the UW-solution, several attempts have been made to identify which components are essential for prevention and reversion of cell swelling in order to create a simplified version of UW-solution [71, 73]. It was found that hydroxyethyl starch, lactobionate and raffinose are the most important compounds for successful purification of human and pig islets [139–141].

4.5.3 Selection of Density Gradient Media

Numerous density gradient media have been assessed for their suitability to separate islets from non-islet tissue [123]. The characteristics and efficiency of density gradient media are determined by viscosity, intrinsic density, osmolality and pH. The finding that Ficoll is contaminated by high concentrations of ions and osmotically active sucrose polymers [142] stimulated the development of iodinated density gradient media that are non-ionic, iso-osmotic and metabolically inert thus improving the posttransplant function of purified islets [143]. The ionidation of media provides a high density without increasing the osmolarity and viscosity. While the latter variable correlates with centrifugation time, the former determines the buoyant density of the separated cell populations reducing the intracellular water content by means of osmotic pressure [144, 145]. It can not be excluded that the pH of a density gradient medium also influences the volume of cells [146].

Currently, isolated and porcine islets are purified utilizing either Ficoll-sodium-diatrizoate (Biocoll) [77, 96] or iodixanol [147, 148] as density gradient media. Apart from a higher efficiency to recover islets after density gradient centrifugation [149], iodixanol has the advantage to reduce cytokine and chemokine release from purified human islets when compared to Biocoll. This feature facilitates higher islet survival during pre-transplant culture [150]. In contrast to human islets which can most efficiently be purified when density gradient media are adjusted to a high osmolarity ranging between 400 and 500 mosm/L [151, 152], bovine, canine and porcine islets can successfully be separated from exocrine tissue by means of iso-osmotic density gradient media [143]. Recent studies have established a nearly iso-osmotic UW-Biocoll density gradient for the purification of human islets that seems to have a larger purification capacity than the Biocoll density gradient [153, 154]. However, according to the best of our knowledge this UW-Biocoll gradient has not been tested for the purification of isolated porcine islets yet.

4.5.4 Density Gradient Centrifugation Technique

Compared with rodents, purification of large volumes of digested tissue, as obtained after human or porcine pancreas digestion, is time consuming when performed in tubes, flasks or other centrifuge vessels [155]. In order to optimize and accelerate the purification of human islets, the Cobe 2991 cell processor was established as centrifugation device for processing large tissue volumes [156]. Since its introduction in 1989 the Cobe 2991 has been established worldwide as standard device for human islet purification [97]. Moreover, the feature of the Cobe 2991 to enable loading of different density gradient media during centrifugal spinning, offered the option to manufacture large-scale linear continuous density gradients which additionally expanded the capacity to purify large volumes of tissue in one single run [157]. Unfortunately, the Cobe 2991 was originally designed to process blood cells at ambient temperature. If cooling during islet density gradient centrifugation is intended, additional significant investments for a cooled room or a custom-made cooling device have to be made [158, 159]. However, whether these technical modifications of the Cobe 2991 are essentially required for successful purification of pig islets can be questioned. In accordance with Chadwick et al. we found that pig islets can be successfully separated from exocrine tissue at room temperature [23, 160].

With respect to the high acquisition costs for a Cobe 2991, attempts were undertaken to re-establish density gradient centrifugation performed in individually loaded flat-bottom flasks [161]. Using this technique for porcine islet purification it was found that the utilization of the Cobe 2991 resulted in significantly higher islet fragmentation and lower post-purification recovery when compared to centrifugation in top-loaded flat-bottom flasks. These observations were related to the higher shearing stress being present in the Cobe 2991 [162]. Nevertheless, the data of Shimoda et al. are in contradiction to findings we previously made in the pig. Although we found an identical islet recovery after purification, the purity estimated by visual assessment, measured by amylase recovery or islet volume-to-DNA ratio was significantly higher after pig islet purification in the Cobe 2991 [37].

In this context it should be stressed that Shimoda et al. utilized a mixed discontinuous-continuous iodixanol density gradient while we have applied neutral density separation as a simple but effective and reproducible approach to purify pig islets according to their buoyancy. To perform neutral density separation which is centrifugation on one layer of iso-osmolar density gradient medium, pig islets are dispensed in 400 mL of cold iso-osmolar Biocoll of a density of 1.082 g/L which is loaded by gravity into a non-activated Cobe 2991. During spinning at 800 × g a volume of 150 mL culture medium is top-layered onto the Biocoll and centrifuged for 5 min at room temperature. Subsequently, 30 mL-fractions are collected in transparent 50 mL-culture flasks for visual inspection by an inverted phase contrast microscope. Purified islets are pooled, washed and suspended in supplemented culture medium [77, 143].

4.5.5 Isokinetic Islet Purification

Although neutral density purification can efficiently separate exocrine tissue from isolated pig islets, ductal, vascular and lymphoid tissue can still contaminate the final islet preparation leading to enhanced rejection of the islet graft [109]. To remove non-islet tissue of a similar density to isolated islets, it was necessary to introduce an additional centrifugation step subsequent to density gradient purification. To facilitate separation of particles according to size a linear isokinetic Biocoll gradient was procuded in tubes in order to enable sedimentation at a constant and slow velocity [163]. After spinning at low speed for 90 s islets were collected from a discrete zone that could be visually distinguished from non-islet tissue fragments. Islets subjected to additional isokinetic centrifugation survived significantly longer in non-immunosuppressed C57/Bl6j mice when compared to conventionally purified pig islets [164].

4.6 Porcine Islet Culture

The primary aim of islet culture is to facilitate survival of isolated islets in vitro in order to pool islet preparations from different donors to provide the critical islet mass for successful xenotransplantation into patients with type 1 diabetes and to enable islet quality assessment prior to transplantation. The present review will focus on islet free-floating culture and is not considering approaches to enhance survival by embedding islets in a matrix or to reduce islet immunogenicity by different culture techniques.

4.6.1 Selecting the Most Suitable Culture Medium

Most of the commercially available media were developed to support expansion and viability of different cell lines [165]. As isolated islets are not proliferating any comparison of different culture media is difficult and has to focus on recovery, viability and function of initially incubated islets. Only very few attempts have been undertaken to identify a medium that is most suitable for successful islet culture. We are aware about only two prospective studies which compare the efficiency of different commercially available culture media to promote function of cultured pig islets. However, no data regarding islet survival were provided in these studies. In one of these studies, Ham's F12 was found being most suitable for pig islet function when compared to CMRL 1066, RPMI 1640 and TCM 199 [166]. The second study identified TCM 199 as the only medium that maintains glucose responsiveness of freshly isolated pig islets during culture because of its content of ATP, AMP and xanthine as precursor for ATP synthesis [49]. The assumption that the complete loss of peripheral alpha cells during pancreas digestion is the main cause for the non-responsiveness of pig islets toward glucose was confirmed in a subsequent study. Supplementation of CMRL 1066 with glucagon re-established glucose-stimulated insulin release of cultured pig islets [50]. These observations clearly underline that the preservation of morphological integrity during enzymatic pancreas digestion is of significant importance for pig islet function.

4.6.2 Glucose Concentration

Because of the complexity of culture media it is difficult to identify the beneficial compounds that made Ham's F12 or TCM 199 to the most efficient media for pig islets. The substance that represent the most important fuel for islet metabolism is glucose. In a prospective study we compared islet survival and in vitro function of pig islets cultured in CMRL 1066 supplemented with different glucose concentrations. It was found that a glucose concentration of 11.1 mmol/L increases pig islet recovery after 8–10 days of culture but reduces islet viability and insulin secretory capacity compared to 5.5 mmol/L glucose [167].

4.6.3 Serum Content

Serum is a complex and abundant source of vitamins, minerals, amino acids, lipids, hormones, enzymes and growth factors. Serum can be collected from bovine, equine, human and porcine sources and is characterized by a significant batch- and species-dependent heterogenity [165]. While a proportion of 10 % fetal calf serum is the standard supplementation for culture of rodent islets, serum has mostly been banned for clinical purposes to prevent transmission of pathogens [168]. Nevertheless, recent studies demonstrate that serum is an essential nutritive supplement to optimize survival and function of human islets particularly during prolonged periods of culture [169, 170]. This observation seems to be even more relevant for cultured pig islets which are characterized by a significant and rapid loss within a short period of time when serum-free medium for clinical purposes is used [171]. A previous study also found that certain serum replacement products do not have the same potency of serum to preserve the integrity of pig islets during culture [172]. Moreover, because of their demanding nutritive requirements long-term cultured pig islets seem to prefer the supplementation of homologous porcine serum when compared to xenogeneic bovine serum [173, 174]. In order to further optimise pig islet survival during culture for research purposes we increased the proportion of porcine serum in culture medium to 20 % [77].

4.6.4 Glutamine Supplementation

Another culture supplement that is used for nearly all cultured cells and tissues is L-glutamine [165]. This essential amino acid does not only serve as precursor for protein synthesis but is playing an important role as major fuel to cover the basal energy consumption of islets [175]. L-glutamine has a sparing effect on utilization of other endogenous nutrients [176] without increasing the oxygen consumption and without stimulating the insulin release [177, 178]. Because the synthesis of proinsulin acccounts for a substantial proportion of islet energy consumption, L-glutamine contributes significantly to save and maintain the endogenous energy stores of cultured islets [179].

Unfortunately, the stability of L-glutamine is low. Depending on the conditions such as temperature or pH, the non-enzymatic degradation of L-glutamine in cell-empty medium can exceed 50 % within 4 days of storage at 37 °C [180, 181]. If proliferating cell lines are present the loss of L-glutamine can reach 90 % within 4 days of culture [182]. For that reason stable glutamine compounds were introduced as supplements for culture media. However, our studies revealed a significant drop in the recovery of cultured pig islets when L-glutamine was replaced by the stable glutamine compound N-acetyl-L-alanyl-L--glutamine. Moreover, pig islets precultured in free L-glutamine were significantly more resistant toward treatment with proinflammatory mediators than islets precultured in the stable glutamine compound [183]. Because of its beneficial effects on cultured pig islets the concentration of L-glutamine has been increased to 5 mmol/L in our current culture protocol. Previously, several companies replaced N-acetyl-L-alanyl-L-glutamine by adding L-alanyl-L-glutamine or L-glycyl-L-glutamine as stable glutamine supplements. So far, no studies have been undertaken to compare the suitability of these dipeptides with free L-glutamine for human or porcine islet culture.

4.6.5 Culture Temperature

The culture temperature is the main determinant for the metabolic demand of cultured cells. According to the Q10 temperature coefficient any reduction of the environmental temperature by 10 °C reduces the metabolic rate of organisms and cells by a factor of approximately three or more [184]. The reduced metabolic activity of pig islets long-term cultured at a temperature of 22 °C is reflected by a two- to fivefold lower basal insulin production and by a two- to threefold decreased insulin stimulatory capacity when compared to islets cultured at 37 °C [174]. Because hypothermia mainly affects the mitochondrial pathways of glucose breakdown the demand for oxygen is also strongly reduced [185]. As a consequence, lower temperatures significantly decrease the extent of central necrosis which correlates with a higher pig islet recovery [174]. A further reduction of the culture temperature to 4 or 1 °C seems to extend the survival rate of pig islets additionally when compared to culture at 22 °C [186].

Nevertheless, in spite of improved recovery, pig islets cultured at 22 °C are characterized by a significantly lower viability and peripheral disintegration of islet morphology [174] which is in agreement with observations in hamster islets cultured at 37 °C or 24 °C [187]. From these observations the question arises whether the higher recovery of islets after 22 °C-culture reflects mummification of dying islets rather than preservation of fully functional and viable islets. In fact, the Leiden group clearly demonstrated that an aliquot of freshly isolated pig islets failed to reverse hyperglycemia in diabetic nude mice while the small proportion of islets surviving after 1 or 2 weeks of culture at 37 °C completely restored normoglycemia in the recipients [188].

4.7 Summary

This overview has highlighted that the optimisation of currently available enzyme blends facilitates the successful isolation of islets from the pancreas of adult donor pigs. The data of the reviewed studies seem to favour a less traumatic dissociation of the pancreas particularly in younger pigs in order to maintain the morphological integrity of released islets. Pig islets can be efficiently purified using iso-osmotic ionidated density gradient media. Morphological integrity is an essential prerequisite to culture pig islets for several days to perform islet quality assessment and to pool islet preparations from different donors to provide the critical islet mass for successful xenotransplantation into patients with type 1 diabetes. The preservation of the morphological and functional integrity of isolated pig islets during culture seems to be supported by the supplementation of culture media with a large proportion of pig serum, a glucose concentration higher than 5.5 mmol/L and a high concentration of L-glutamine.

However, apart from the discussion of the technical details and prerequisites for a successful isolation of pig islets the selection of a suitable pig strain is of overwhelming importance for success and failure of any single step within the procedure of pancreas processing for islet isolation.

References

1. Cooper DK, Ekser B, Tector AJ. A brief history of clinical xenotransplantation. Int J Surg. 2015. doi:10.1016/j.ijsu.2015.06.060.
2. Ekser B, Cooper DK, Tector AJ. The need for xenotransplantation as a source of organs and cells for clinical transplantation. Int J Surg. 2015. doi:10.1016/j.ijsu.2015.06.066.
3. Ibrahim Z, Busch J, Awwad M, Wagner R, Wells K, Cooper DK. Selected physiologic compatibilities and incompatibilities between human and porcine organ systems. Xenotransplantation. 2006;13(6):488–99. doi:10.1111/j.1399-3089.2006.00346.x.
4. Wijkstrom M, Bottino R, Iwase H, Hara H, Ekser B, van der Windt D, et al. Glucose metabolism in pigs expressing human genes under an insulin promoter. Xenotransplantation. 2015;22(1):70–9. doi:10.1111/xen.12145.
5. Groth CG, Korsgren O, Tibell A, Tollemar J, Moller E, Bolinder J, et al. Transplantation of porcine fetal pancreas to diabetic patients. Lancet. 1994;344(8934):1402–4.
6. Elliott RB, Escobar L, Tan PL, Muzina M, Zwain S, Buchanan C. Live encapsulated porcine islets from a type 1 diabetic patient 9.5 yr after xenotransplantation. Xenotransplantation. 2007;14(2):157–61. doi:10.1111/j.1399-3089.2007.00384.x. doi:XEN384 [pii].
7. Cooper DK, Ekser B, Tector AJ. Immunobiological barriers to xenotransplantation. Int J Surg. 2015. doi:10.1016/j.ijsu.2015.06.068.
8. Cardona K, Korbutt GS, Milas Z, Lyon J, Cano J, Jiang W, et al. Long-term survival of neonatal porcine islets in nonhuman primates by targeting costimulation pathways. Nat Med. 2006;12(3):304–6.
9. Hering BJ, Wijkstrom M, Graham ML, Hardstedt M, Aasheim TC, Jie T, et al. Prolonged diabetes reversal after intraportal xenotransplantation of wild-type porcine islets in immunosuppressed nonhuman primates. Nat Med. 2006;12(3):301–3.
10. Goto M, Tjernberg J, Dufrane D, Elgue G, Brandhorst D, Ekdahl KN, et al. Dissecting the instant blood-mediated inflammatory reaction in islet xenotransplantation. Xenotransplantation. 2008;15(4):225–34.
11. Bottino R, Wijkstrom M, van der Windt DJ, Hara H, Ezzelarab M, Murase N, et al. Pig-to-monkey islet xenotransplantation using multi-transgenic pigs. Am J Transplant. 2014;14(10):2275–87. doi:10.1111/ajt.12868.
12. Chen Y, Stewart JM, Gunthart M, Hawthorne WJ, Salvaris EJ, O'Connell PJ, et al. Xenoantibody response to porcine islet cell transplantation using GTKO, CD55, CD59, and fucosyltransferase multiple transgenic donors. Xenotransplantation. 2014;21(3):244–53. doi:10.1111/xen.12091.
13. Hawthorne WJ, Salvaris EJ, Phillips P, Hawkes J, Liuwantara D, Burns H, et al. Control of IBMIR in neonatal porcine islet xenotransplantation in baboons. Am J Transplant. 2014;14(6):1300–9. doi:10.1111/ajt.12722.
14. Ellis CE, Korbutt GS. Justifying clinical trials for porcine islet xenotransplantation. Xenotransplantation. 2015;22(5):336–44. doi:10.1111/xen.12196.
15. Denner J, Schuurman HJ, Patience C. The International Xenotransplantation Association consensus statement on conditions for undertaking clinical trials of porcine islet products in type 1 diabetes – chapter 5: strategies to prevent transmission of porcine endogenous retroviruses. Xenotransplantation. 2009;16(4):239–48. doi:10.1111/j.1399-3089.2009.00544.x. doi:XEN544 [pii].

16. Beckwith J, Nyman JA, Flanagan B, Schrover R, Schuurman HJ. A health-economic analysis of porcine islet xenotransplantation. Xenotransplantation. 2010;17(3):233–42. doi:10.1111/j.1399-3089.2010.00586.x. doi:XEN586 [pii].
17. Korbutt GS. What type of islets should be used? Xenotransplantation. 2008;15(2):81–2.
18. Nagaraju S, Bottino R, Wijkstrom M, Trucco M, Cooper DK. Islet xenotransplantation: what is the optimal age of the islet-source pig? Xenotransplantation. 2015;22(1):7–19. doi:10.1111/xen.12130.
19. Krishnan R, Buder B, Alexander M, Foster 3rd CE, Lakey JR. Juvenile porcine islets can restore euglycemia in diabetic athymic nude mice after xenotransplantation. Transplantation. 2015;99(4):710–6. doi:10.1097/TP.0000000000000667.
20. Kirchhof N, Hering BJ, Geiss V, Federlin K, Bretzel RG. Evidence for breed-dependent differences in porcine islets of Langerhans. Transplant Proc. 1994;26(2):616–7.
21. Ulrichs K, Bosse M, Heiser A, Eckstein V, Wacker HH, Thiede A, et al. Histomorphological characteristics of the porcine pancreas as a basis for the isolation of islets of Langerhans. Xenotransplantation. 1995;2:176–87.
22. Meyer T, Czub S, Chodnewska I, Beutner U, Hamelmann W, Klock G, et al. Expression pattern of extracellular matrix proteins in the pancreas of various domestic pig breeds, the Goettingen Minipig and the Wild Boar. Ann Transplant. 1997;2(3):17–26.
23. Brandhorst H. Isolation, purification and transplantation of porcine islets of Langerhans considering culture and cryopreservation of isolated islets. Thesis, Justus-Liebig-University, Giessen. 1992.
24. Heiser A, Ulrichs K, Muller-Ruchholtz W. Influence of porcine strain, age, and pH of the isolation medium on porcine pancreatic islet isolation success. Transplant Proc. 1994;26(2):618–20.
25. British-Pig-Association. The British landrace. 2015. Available at: http://www.britishpigs.org.uk/breed_la
26. Canario L, Pere MC, Tribout T, Thomas F, David C, Gogue J, et al. Estimation of genetic trends from 1977 to 1998 of body composition and physiological state of Large White pigs at birth. Anim Int J Anim Biosci. 2007;1(10):1409–13. doi:10.1017/S1751731107000766.
27. Hocquette JF, Gondret F, Baeza E, Medale F, Jurie C, Pethick DW. Intramuscular fat content in meat-producing animals: development, genetic and nutritional control, and identification of putative markers. Anim Int J Anim Biosci. 2010;4(2):303–19. doi:10.1017/S1751731109991091.
28. Althen TG, Gerrits RJ. Pituitary and serum growth hormone levels in Duroc and Yorkshire swine genetically selected for high and low backfat. J Anim Sci. 1976;42(6):1490–7.
29. Kasser TR, Martin RJ, Gahagan JH, Wangsness PJ. Fasting plasma hormones and metabolites in feral and domestic newborn pigs. J Anim Sci. 1981;53(2):420–6.
30. Hoffman EC, Wangsness PJ, Hagen DR, Etherton TD. Fetuses of lean and obese swine in late gestation: body composition, plasma hormones and muscle development. J Anim Sci. 1983;57(3):609–20.
31. Klindt J, Buonomo FC, Yen JT, Pond WG, Mersmann HJ. Administration of porcine somatotropin by daily injection: growth and endocrine responses in genetically lean and obese barrows and gilts. J Anim Sci. 1995;73(11):3294–303.
32. Kim JH, Kim HI, Lee KW, Yu JE, Kim SH, Park HS, et al. Influence of strain and age differences on the yields of porcine islet isolation: extremely high islet yields from SPF CMS miniature pigs. Xenotransplantation. 2007;14(1):60–6.
33. Ricordi C, Finke EH, Lacy PE. A method for the mass isolation of islets from the adult pig pancreas. Diabetes. 1986;35(6):649–53.
34. van Deijnen JH, Hulstaert CE, Wolters GH, van Schilfgaarde R. Significance of the peri-insular extracellular matrix for islet isolation from the pancreas of rat, dog, pig, and man. Cell Tissue Res. 1992;267(1):139–46.
35. Van Deijnen JH, Van Suylichem PT, Wolters GH, Van Schilfgaarde R. Distribution of collagens type I, type III and type V in the pancreas of rat, dog, pig and man. Cell Tissue Res. 1994;277(1):115–21.
36. Cabrera O, Berman DM, Kenyon NS, Ricordi C, Berggren PO, Caicedo A. The unique cytoarchitecture of human pancreatic islets has implications for islet cell function. Proc Natl Acad Sci U S A. 2006;103(7):2334–9.
37. Brandhorst D, Brandhorst H, Hering BJ, Federlin K, Bretzel RG. Islet isolation from the pancreas of large mammals and humans: 10 years of experience. Exp Clin Endocrinol Diabetes. 1995;103:3–14.
38. Socci C, Ricordi C, Davalli AM, Staudacher C, Baro P, Vertova A, et al. Selection of donors significantly improves pig islet isolation yield. Horm Metab Res Suppl. 1990;25:32–4.
39. Krickhahn M, Meyer T, Buhler C, Thiede A, Ulrichs K. Highly efficient isolation of porcine islets of Langerhans for xenotransplantation: numbers, purity, yield and in vitro function. Ann Transplant. 2001;6(3):48–54.
40. Anazawa T, Balamurugan AN, Matsumoto S, Lafreniere SA, O'Brien TD, Sutherland DE, et al. Rapid quantitative assessment of the pig pancreas biopsy predicts islet yield. Transplant Proc. 2010;42(6):2036–9. doi:10.1016/j.transproceed.2010.05.113. doi:S0041-1345(10)00778-5 [pii].
41. Bottino R, Balamurugan AN, Smetanka C, Bertera S, He J, Rood PP, et al. Isolation outcome and functional characteristics of young and adult pig pancre-

atic islets for transplantation studies. Xenotransplantation. 2007;14(1):74–82.
42. Dufrane D, Goebbels RM, Fdilat I, Guiot Y, Gianello P. Impact of porcine islet size on cellular structure and engraftment after transplantation: adult versus young pigs. Pancreas. 2005;30(2):138–47. doi:00006676-200503000-00007 [pii].
43. Krickhahn M, Buhler C, Meyer T, Thiede A, Ulrichs K. The morphology of islets within the porcine donor pancreas determines the isolation result: successful isolation of pancreatic islets can now be achieved from young market pigs. Cell Transplant. 2002;11(8):827–38.
44. Hilling D, Rijkelijkhuizen JK, Tons HA, Terpstra OT, Bouwman E. Amount and distribution of collagen in the pancreas have no effect on porcine islet isolation outcome. Transplant Proc. 2009;41(1):326–7. doi:10.1016/j.transproceed.2008.10.065. doi:S0041-1345(08)01573-X [pii].
45. van Suylichem PT, van Deijnen JE, Wolters GH, van Schilfgaarde R. Amount and distribution of collagen in pancreatic tissue of different species in the perspective of islet isolation procedures. Cell Transplant. 1995;4(6):609–14.
46. White SA, Hughes DP, Contractor HH, London NJ. An investigation into the distribution of different collagen types within adult and juvenile porcine pancreata. J Mol Med. 1999;77(1):79–82.
47. Aumailley M, von der Mark H, Timpl R. Size and domain structure of collagen VI produced by cultured fibroblasts. FEBS Lett. 1985;182(2):499–502.
48. Marchetti P, Scharp DW, Finke EH, Giannarelli R, Swanson CJ, Olack BJ, et al. Factors affecting in vitro and in vivo function of isolated porcine islets. Transplant Proc. 1994;26(6):3419–20.
49. Davalli AM, Bertuzzi F, Socci C, Scaglia L, Gavazzi F, Freschi M, et al. Paradoxical release of insulin by adult pig islets in vitro. Recovery after culture in a defined tissue culture medium. Transplantation. 1993;56(1):148–54.
50. Bertuzzi F, Berra C, Socci C, Davalli AM, Calori G, Freschi M, et al. Glucagon improves insulin secretion from pig islets in vitro. J Endocrinol. 1995;147(1):87–93.
51. Ricordi C, Socci C, Davalli AM, Staudacher C, Baro P, Vertova A, et al. Effect of pancreas retrieval procedure on islet isolation in the swine. Transplant Proc. 1990;22(2):442–3.
52. Warnock GL, Katyal D, Okamura J, Lakey JRT, Rajotte RV. Studies of the isolation, viability, and preservation of purified islets after surgical pancreatectomy in large pigs. Xenotransplantation. 1995;2:161–4.
53. Kerr-Conte J, Pattou F, Hober C, Ple A, Riquier M, Proye C, et al. Induction of cerebral death in slaughterhouse pigs facilitates pancreas harvesting for islet isolation. Transplant Proc. 1994;26(2):614–5.
54. Meyer T, Buhler C, Debus S, Beutner U, Otto C, Timmermann W, et al. Traumatized pigs are unsuitable as organ donors for pancreatic islet isolation. Transplant Proc. 1998;30(5):2475–7.
55. Wright MJ, Cavanagh TJ, Fetterhoff TJ, Wile KJ. Effect of blood content on porcine pancreatic dissociation and islet yield. Transplant Proc. 1994;26(6):3442.
56. Dufrane D, D'Hoore W, Goebbels RM, Saliez A, Guiot Y, Gianello P. Parameters favouring successful adult pig islet isolations for xenotransplantation in pig-to-primate models. Xenotransplantation. 2006;13(3):204–14.
57. Anazawa T, Balamurugan AN, Papas KK, Avgoustiniatos ES, Ferrer J, Matsumoto S, et al. Improved method of porcine pancreas procurement with arterial flush and ductal injection enhances islet isolation outcome. Transplant Proc. 2010;42(6):2032–5. doi:10.1016/j.transproceed.2010.05.110. doi:S0041-1345(10)00775-X [pii].
58. Warnock GL, Ellis D, Rajotte RV, Dawidson I, Baekkeskov S, Egebjerg J. Studies of the isolation and viability of human islets of Langerhans. Transplantation. 1988;45(5):957–63.
59. Kin T, Johnson PR, Shapiro AM, Lakey JR. Factors influencing the collagenase digestion phase of human islet isolation. Transplantation. 2007;83(1):7–12.
60. Lakey JR, Warnock GL, Shapiro AM, Korbutt GS, Ao Z, Kneteman NM, et al. Intraductal collagenase delivery into the human pancreas using syringe loading or controlled perfusion. Cell Transplant. 1999;8(3):285–92.
61. Hilling DE, Rijkelijkhuizen JK, Marang-van de Mheen PJ, Tons A, Terpstra OT, Bouwman E. Morphological changes of porcine islets of Langerhans after collagenase and HBSS infusion of the pancreas. Xenotransplantation. 2010;17(6):413–7.
62. Johnson PRV, Chen R, Smith JS, Wheatley L, James RFL, London NJM. Influence of different loading pressures on the distribution of collagenase within the porcine pancreas. Horm Metab Res. 1996;28:55.
63. Lakey JR, Rajotte RV, Warnock GL, Kneteman NM. Human pancreas preservation prior to islet isolation. Cold ischemic tolerance. Transplantation. 1995;59(5):689–94.
64. Brandhorst D, Iken M, Tanioka Y, Brendel MD, Bretzel RG, Brandhorst H. Influence of collagenase loading on long-term preservation of pig pancreas by the two-layer method for subsequent islet isolation. Transplantation. 2005;79(1):38–43.
65. Sato S, Ohgawara H, Katagiri N, Nakagawa Y, Aikawa E, Omori Y, et al. A simple method for purification of adult pig pancreatic endocrine cells. Pancreas. 1997;14(4):400–6.
66. Socci C, Davalli AM, Vignali A, Pontiroli AE, Maffi P, Magistretti P, et al. A significant increase of islet yield by early injection of collagenase into the pancreatic duct of young donors. Transplantation. 1993;55(3):661–3.

67. Shimoda M, Itoh T, Sugimoto K, Iwahashi S, Takita M, Chujo D, et al. Improvement of collagenase distribution with the ductal preservation for human islet isolation. Islets. 2012;4(2):130–7. doi:10.4161/isl.19255.
68. Mellert J, Hering BJ, Hopt UT, Bretzel RG, Hufnagel B, Pfeffer F, et al. Exchange of pancreata and islets between centers for experimental islet transplantation in the pig. Transplant Proc. 1991;23(5):2435–6.
69. van der Burg MP, Basir I, Zwaan RP, Bouwman E. Porcine islet preservation during isolation in University of Wisconsin solution. Transplant Proc. 1998;30(2):360–1.
70. Heald KA, Jay TR, Topham D, Downing R. The influence of collagenase solvent on the isolation of islets from 5 week old pigs: a comparison of TCM-199 and UW. J Mol Med. 1999;77(1):83–6.
71. White SA, Contractor HH, Hughes DP, Johnson PR, Clayton HA, Bell PR, et al. Influence of different collagenase solvents and timing of their delivery on porcine islet isolation. Br J Surg. 1996;83(10):1350–5.
72. Johnson PR, Button CL, Roberts DL, Contractor HH, White SW, Bell PR, et al. University of Wisconsin solution inhibits the class II collagenase within crude Clostridium histolyticum collagenase. Transplant Proc. 1995;27(6):3286.
73. O'Neil JJ, Stegemann JP, Nicholson DT, Gagnon KA, Solomon BA, Mullon CJ. The isolation and function of porcine islets from market weight pigs. Cell Transplant. 2001;10(3):235–46.
74. Contractor HH, Johnson PR, Chadwick DR, Robertson GS, London NJ. The effect of UW solution and its components on the collagenase digestion of human and porcine pancreas. Cell Transplant. 1995;4(6):615–9.
75. Mirenda V, Le Mauff B, Huvelin JM, Sigalla J, Auget JL, Soulillou JP. Improvement of adult porcine islet recovery by early collagenase injection and low temperature digestion. Transplant Proc. 1997;29(4):2265.
76. Johnson PR, White SA, London NJ. Collagenase and human islet isolation. Cell Transplant. 1996;5(4):437–52.
77. Brandhorst H, Brandhorst D, Hering BJ, Bretzel RG. Significant progress in porcine islet mass isolation utilizing liberase HI for enzymatic low-temperature pancreas digestion. Transplantation. 1999;68(3):355–61.
78. Dono K, Gotoh M, Monden M, Kanai T, Fukuzaki T, Mori T. Low temperature collagenase digestion for islet isolation from 48-hour cold-preserved rat pancreas. Transplantation. 1994;57(1):22–6.
79. Goto T, Tanioka Y, Sakai T, Terai S, Kamoda Y, Li S, et al. Application of the two-layer method on pancreas digestion results in improved islet yield and maintained viability of isolated islets. Transplantation. 2007;83(6):754–8. doi:10.1097/01.tp.0000256338.53305.a9. 00007890-200703270-00017 [pii].
80. Niederau C, Fronhoffs K, Klonowski H, Schulz HU. Active pancreatic digestive enzymes show striking differences in their potential to damage isolated rat pancreatic acinar cells. J Lab Clin Med. 1995;125(2):265–75.
81. Schwarz A, Jung M, Heeckt P, Cammerer G, Popp J, Nussler AK, et al. Optimizing of isolation, purification and function of islets from the pancreas of the adult large German pig. Transplant Proc. 1997;29(4):1965–6.
82. White SA, Djaballah H, Hughes DP, Roberts DL, Contractor HH, Pathak S, et al. A preliminary study of the activation of endogenous pancreatic exocrine enzymes during automated porcine islet isolation. Cell Transplant. 1999;8(3):265–76.
83. Heiser A, Ulrichs K, Muller-Ruchholtz W. Prophylactic trypsin inhibition during the isolation procedure guarantees reproducible, high porcine islet yields. Xenotransplantation. 1994;1:66–8.
84. Basir I, van der Burg MP, Scheringa M, Tons A, Bouwman E. Improved outcome of pig islet isolation by Pefabloc inhibition of trypsin. Transplant Proc. 1997;29(4):1939–41.
85. Bai RX, Fujimori K, Koja S, Sekiguchi S, Doi H, Tsukamoto S, et al. Effect of prophylactic administration of trypsin inhibitors in porcine pancreas islet isolation. Transplant Proc. 1998;30(2):349–52.
86. Shimoda M, Noguchi H, Fujita Y, Takita M, Ikemoto T, Chujo D, et al. Improvement of porcine islet isolation by inhibition of trypsin activity during pancreas preservation and digestion using alpha1-antitrypsin. Cell Transplant. 2012;21(2–3):465–71. doi:10.3727/096368911X605376.
87. Wolters GH, Vos-Scheperkeuter GH, van Deijnen JH, van Schilfgaarde R. An analysis of the role of collagenase and protease in the enzymatic dissociation of the rat pancreas for islet isolation. Diabetologia. 1992;35(8):735–42.
88. Brandhorst H, Brendel MD, Eckhard M, Bretzel RG, Brandhorst D. Influence of neutral protease activity on human islet isolation outcome. Transplant Proc. 2005;37(1):241–2.
89. Balamurugan AN, Loganathan G, Bellin MD, Wilhelm JJ, Harmon J, Anazawa T, et al. A new enzyme mixture to increase the yield and transplant rate of autologous and allogeneic human islet products. Transplantation. 2012;93(7):693–702. doi:10.1097/TP.0b013e318247281b.
90. Cavanagh TJ, Lakey JR, Dwulet F, Wright MJ, Wile K, Albertson T, et al. Improved pig islet yield and post-culture recovery using Liberase PI purified enzyme blend. Transplant Proc. 1998;30(2):367.
91. Josefsen K, Stenvang JP, Kindmark H, Berggren PO, Horn T, Kjaer T, et al. Fluorescence-activated cell sorted rat islet cells and studies of the insulin secretory process. J Endocrinol. 1996;149(1):145–54.

92. Stenn KS, Link R, Moellmann G, Madri J, Kuklinska E. Dispase, a neutral protease from Bacillus polymyxa, is a powerful fibronectinase and type IV collagenase. J Invest Dermatol. 1989;93(2):287–90.
93. Cheng JY, Raghunath M, Whitelock J, Poole-Warren L. Matrix components and scaffolds for sustained islet function. Tissue Eng Part B Rev. 2011;17(4):235–47. doi:10.1089/ten.TEB.2011.0004.
94. Irving-Rodgers HF, Choong FJ, Hummitzsch K, Parish CR, Rodgers RJ, Simeonovic CJ. Pancreatic islet basement membrane loss and remodeling after mouse islet isolation and transplantation: impact for allograft rejection. Cell Transplant. 2014;23(1):59–72. doi:10.3727/096368912X659880.
95. Brandhorst H, Iken M, Scott 3rd WE, Papas KK, Theisinger B, Johnson PR, et al. Quality of isolated pig islets is improved using perfluorohexyloctane for pancreas storage in a split lobe model. Cell Transplant. 2013;22(8):1477–83. doi:10.3727/0963 68912X657639.
96. Brandhorst H, Theisinger B, Guenther B, Johnson PR, Brandhorst D. Pancreatic L-glutamine administration protects pig islets from cold ischemic injury and increases resistance toward inflammatory mediators. Cell Transplant. 2015. doi:10.3727/0963689 15X688623. online.
97. Lakey JR, Burridge PW, Shapiro AM. Technical aspects of islet preparation and transplantation. Transpl Int. 2003;16(9):613–32.
98. Ricordi C, Lacy PE, Finke EH, Olack BJ, Scharp DW. Automated method for isolation of human pancreatic islets. Diabetes. 1988;37(4):413–20.
99. Gray DW, McShane P, Grant A, Morris PJ. A method for isolation of islets of Langerhans from the human pancreas. Diabetes. 1984;33(11):1055–61.
100. Ricordi C, Socci C, Davalli AM, Staudacher C, Baro P, Vertova A, et al. Isolation of the elusive pig islet. Surgery. 1990;107(6):688–94.
101. Balamurugan AN, Chang Y, Fung JJ, Trucco M, Bottino R. Flexible management of enzymatic digestion improves human islet isolation outcome from sub-optimal donor pancreata. Am J Transplant. 2003;3(9):1135–42.
102. Toomey P, Chadwick DR, Contractor H, Bell PR, James RF, London NJ. Porcine islet isolation: prospective comparison of automated and manual methods of pancreatic collagenase digestion. Br J Surg. 1993;80(2):240–3.
103. Scharp DW, Murphy JJ, Newton WT, Ballinger WF, Lacy PE. Transplantation of islets of Langerhans in diabetic rhesus monkeys. Surgery. 1975;77(1):100–5.
104. Traverso LW, Gomez RR. Hemodynamic measurements after administration of aprotinin and/or heparin during pancreatic cell autotransplantation in the dog, pig, and monkey. Ann Surg. 1982;195(4):479–85.
105. Casey JJ, Lakey JR, Ryan EA, Paty BW, Owen R, O'Kelly K, et al. Portal venous pressure changes after sequential clinical islet transplantation. Transplantation. 2002;74(7):913–5.
106. Hyon SH, Ceballos MC, Barbich M, Groppa R, Grosembacher L, Vieiro MM, et al. Effect of the embolization of completely unpurified islets on portal vein pressure and hepatic biochemistry in clinical practice. Cell Transplant. 2004;13(1):61–5.
107. Mittal VK, Toledo-Pereyra LH, Sharma M, Ramaswamy K, Puri VK, Cortez JA, et al. Acute portal hypertension and disseminated intravascular coagulation following pancreatic islet autotransplantation after subtotal pancreatectomy. Transplantation. 1981;31(4):302–4.
108. Toledo-Pereyra LH, Rowlett AL, Cain W, Rosenberg JC, Gordon DA, MacKenzie GH. Hepatic infarction following intraportal islet cell autotransplantation after near-total pancreatectomy. Transplantation. 1984;38(1):88–9.
109. Gotoh M, Maki T, Satomi S, Porter J, Monaco AP. Immunological characteristics of purified pancreatic islet grafts. Transplantation. 1986;42(4):387–90.
110. Clayton HA, Davies JE, Sutton CD, Bell PR, Dennison AR. A coculture model of intrahepatic islet transplantation: activation of Kupffer cells by islets and acinar tissue. Cell Transplant. 2001;10(1):101–8.
111. Kessler L, Jesser C, Belcourt A, Pinget M. Influence of acinar tissue contamination on encapsulated pancreatic islets: morphological and functional studies. Transplantation. 1997;63(10):1537–40.
112. Kin T, Senior P, O'Gorman D, Richer B, Salam A, Shapiro AM. Risk factors for islet loss during culture prior to transplantation. Transpl Int. 2008;21(11):1029–35.
113. Loganathan G, Dawra RK, Pugazhenthi S, Guo Z, Soltani SM, Wiseman A, et al. Insulin degradation by acinar cell proteases creates a dysfunctional environment for human islets before/after transplantation: benefits of alpha-1 antitrypsin treatment. Transplantation. 2011;92(11):1222–30. doi:10.1097/TP.0b013e318237585c.
114. Rajotte RV. Isolation and assessment of islet quality. Xenotransplantation. 2008;15(2):93–5.
115. Korbutt GS. The International Xenotransplantation Association consensus statement on conditions for undertaking clinical trials of porcine islet products in type 1 diabetes – chapter 3: pig islet product manufacturing and release testing. Xenotransplantation. 2009;16(4):223–8. doi:10.1111/j.1399-3089.2009.00542.x. doi:XEN542 [pii].
116. Gray DW, Sutton R, McShane P, Peters M, Morris PJ. Exocrine contamination impairs implantation of pancreatic islets transplanted beneath the kidney capsule. J Surg Res. 1988;45(5):432–42.
117. Downing R, Morrissey S, Kiske D, Scharp DW. Does the purity of intraportal islet isografts affect their endocrine function? J Surg Res. 1986;41(1):41–6.

118. Al-Abdullah IH, Anil Kumar MS, Kelly-Sullivan D, Abouna GM. Site for unpurified islet transplantation is an important parameter for determination of the outcome of graft survival and function. Cell Transplant. 1995;4(3):297–305.
119. Matsumoto S, Takita M, Shimoda M, Sugimoto K, Itoh T, Chujo D, et al. Impact of tissue volume and purification on clinical autologous islet transplantation for the treatment of chronic pancreatitis. Cell Transplant. 2012;21(4):625–32. doi:10.3727/096368911X623899. doi:ct0403matsumoto [pii].
120. Nicolae M, Ketchum RJ, Jahr H, Janoff H, Brayman KL. Islet xenografts in mice and rats: I. Exocrine tissue is detrimental to engraftment and survival of human islet xenografts. Transplant Proc. 1994;26(2):774.
121. Badet L, Titus T, Metzen E, Handa A, McShane P, Chang LW, et al. The interaction between primate blood and mouse islets induces accelerated clotting with islet destruction. Xenotransplantation. 2002;9(2):91–6.
122. van der Windt DJ, Marigliano M, He J, Votyakova TV, Echeverri GJ, Ekser B, et al. Early islet damage after direct exposure of pig islets to blood: has humoral immunity been underestimated? Cell Transplant. 2012;21(8):1791–802. doi:10.3727/096368912X653011.
123. London NJM, James RFL, Bell PRF. Islet purification. In: Ricordi C, editor. Pancreatic islet cell transplantation. Pancreatic islet cell transplantation. Austin: Landes Co; 1992. p. 113–23.
124. Salvalaggio PR, Deng S, Ariyan CE, Millet I, Zawalich WS, Basadonna GP, et al. Islet filtration: a simple and rapid new purification procedure that avoids ficoll and improves islet mass and function. Transplantation. 2002;74(6):877–9. doi:10.1097/01.TP.0000028781.41729.5B.
125. London NJ, Toomey P, Contractor H, Thirdborough ST, James RF, Bell PR. The effect of osmolality and glucose concentration on the purity of human islet isolates. Transplant Proc. 1992;24(3):1002.
126. Caroll PB. Anatomy and physiology of islets of Langerhans. In: Ricordi C, editor. Pancreatic islet cell transplantation. Pancreatic islet cell transplantation. Austin: Landes Co; 1992. p. 7–18.
127. Ichikawa A. Fine structural changes in response to hormonal stimulation of the perfused canine pancreas. J Cell Biol. 1965;24:369–85.
128. Scheele GA, Palade GE. Studies on the guinea pig pancreas. Parallel discharge of exocrine enzyme activities. J Biol Chem. 1975;250(7):2660–70.
129. Scheele G, Tartakoff A. Exit of nonglycosylated secretory proteins from the rough endoplasmic reticulum is asynchronous in the exocrine pancreas. J Biol Chem. 1985;260(2):926–31.
130. Fischer H, Polikarpov I, Craievich AF. Average protein density is a molecular-weight-dependent function. Protein Sci. 2004;13(10):2825–8. doi:10.1110/ps.04688204.
131. DiMagno EP. Human exocrine pancreatic enzyme secretion. In: Go VLW, Gardner JD, Brooks FP, Lebenthal E, DiMagno EP, Scheele GA, editors. The exocrine pancreas: biology, pathobiology, and diseases. The exocrine pancreas: biology, pathobiology, and diseases. New York: Raven; 1986. p. 193–211.
132. Caballero-Corbalan J, Brandhorst H, Malm H, Felldin M, Foss A, Salmela K, et al. Using HTK for prolonged pancreas preservation prior to human islet isolation. J Surg Res. 2012;175(1):163–8. doi:10.1016/j.jss.2011.03.012. doi:S0022-4804(11)00207-1 [pii].
133. Eckhard M, Brendel MD, Brandhorst D, Brandhorst H, Bretzel RG. Can the density of native pancreatic tissue slices predict human islet isolation and purification outcome? Transplant Proc. 2004;36(9):2845–8.
134. Lambotte L. Effect of anoxia and ATP depletion on the membrane potential and permeability of dog liver. J Physiol. 1977;269(1):53–76.
135. Belzer FO, Southard JH. Principles of solid-organ preservation by cold storage. Transplantation. 1988;45(4):673–6.
136. Berthon B, Claret M, Mazet JL, Poggioli J. Volume- and temperature-dependent permeabilities in isolated rat liver cells. J Physiol. 1980;305:267–77.
137. van der Burg MP, Gooszen HG, Ploeg RJ, Guicherit OR, Scherft JP, Terpstra JL, et al. Pancreatic islet isolation with UW solution: a new concept. Transplant Proc. 1990;22(4):2050–1.
138. Robertson GS, Chadwick D, Contractor H, Rose S, Chamberlain R, Clayton H, et al. Storage of human pancreatic digest in University of Wisconsin solution significantly improves subsequent islet purification. Br J Surg. 1992;79(9):899–902.
139. Robertson GS, Chadwick DR, Davies J, Rose S, Contractor H, James RF, et al. The effectiveness of components of University of Wisconsin solution in improving human pancreatic islet purification. Transplantation. 1994;57(3):346–54.
140. Chadwick DR, Robertson GS, Rose S, Contractor H, James RF, Bell PR, et al. Storage of porcine pancreatic digest prior to islet purification. The benefits of UW solution and the roles of its individual components. Transplantation. 1993;56(2):288–93.
141. Chadwick DR, Robertson GS, Contractor HH, Rose S, Johnson PR, James RF, et al. Storage of pancreatic digest before islet purification. The influence of colloids and the sodium to potassium ratio in University of Wisconsin-based preservation solutions. Transplantation. 1994;58(1):99–104.
142. Scharp DW, Kemp CB, Knight MJ, Ballinger WF, Lacy PE. The use of ficoll in the preparation of viable islets of langerhans from the rat pancreas. Transplantation. 1973;16(6):686–9.
143. Hering BJ, Muench KP, Schelz J, Amelang D, Heitfeld M, Bretzel RG, et al. The evaluation of neutral density separation utilizing Ficoll-sodium diatrizoate and Nycodenz and centrifugal elutriation in

the purification of bovine and canine islet preparations. Horm Metab Res Suppl. 1990;25:57–63.
144. Williams N, Kraft N, Shortman K. The separation of different cell classes from lymphoid organs. VI. The effect of osmolarity of gradient media on the density distribution of cells. Immunology. 1972;22(5):885–99.
145. van Suylichem PT, Wolters GH, van Schilfgaarde R. The efficacy of density gradients for islet purification: a comparison of seven density gradients. Transpl Int. 1990;3(3):156–61.
146. Williams N, Shortman K. The separation of different cell classes from lymphoid organs. The effect of pH on the buoyant density of lymphocytes and erythrocytes. Aust J Exp Biol Med Sci. 1972;50(2):133–51.
147. van der Burg MP, Basir I, Bouwman E. No porcine islet loss during density gradient purification in a novel iodixanol in University of Wisconsin solution. Transplant Proc. 1998;30(2):362–3.
148. van der Burg MP, Graham JM. Iodixanol density gradient preparation in university of wisconsin solution for porcine islet purification. TheScientificWorldJOURNAL. 2003;3:1154–9.
149. Noguchi H, Ikemoto T, Naziruddin B, Jackson A, Shimoda M, Fujita Y, et al. Iodixanol-controlled density gradient during islet purification improves recovery rate in human islet isolation. Transplantation. 2009;87(11):1629–35.
150. Mita A, Ricordi C, Messinger S, Miki A, Misawa R, Barker S, et al. Antiproinflammatory effects of iodixanol (OptiPrep)-based density gradient purification on human islet preparations. Cell Transplant. 2010;19(12):1537–46. doi:10.3727/096368910X516600. doi:ct2322mita [pii].
151. Eckhard M, Brandhorst D, Brandhorst H, Brendel MD, Bretzel RG. Optimization in osmolality and range of density of a continuous ficoll-sodium-diatrizoate gradient for isopycnic purification of isolated human islets. Transplant Proc. 2004;36(9):2849–54.
152. Noguchi H, Naziruddin B, Shimoda M, Fujita Y, Chujo D, Takita M, et al. Evaluation of osmolality of density gradient for human islet purification. Cell Transplant. 2012;21(2–3):493–500. doi:10.3727/096368911X605402.
153. Huang GC, Zhao M, Jones P, Persaud S, Ramracheya R, Lobner K, et al. The development of new density gradient media for purifying human islets and islet-quality assessments. Transplantation. 2004;77(1):143–5.
154. Barbaro B, Salehi P, Wang Y, Qi M, Gangemi A, Kuechle J, et al. Improved human pancreatic islet purification with the refined UIC-UB density gradient. Transplantation. 2007;84(9):1200–3. doi:10.1097/01.tp.0000287127.00377.6f.
155. Scharp DW, Lacy PE, Finke E, Olack B. Low-temperature culture of human islets isolated by the distention method and purified with Ficoll or Percoll gradients. Surgery. 1987;102(5):869–79.
156. Lake SP, Bassett PD, Larkins A, Revell J, Walczak K, Chamberlain J, et al. Large-scale purification of human islets utilizing discontinuous albumin gradient on IBM 2991 cell separator. Diabetes. 1989;38 Suppl 1:143–5.
157. Robertson GS, Chadwick DR, Contractor H, James RF, London NJ. The optimization of large-scale density gradient isolation of human islets. Acta Diabetol. 1993;30(2):93–8.
158. Swift S, Kin T, Mirbolooki M, Wilson R, Lakey JR. Comparison of cooling systems during islet purification. Cell Transplant. 2006;15(2):175–80.
159. Adewola A, Mage R, Hansen M, Barbaro B, Mendoza-Elias J, Harvat T, et al. Comparing cooling systems for the COBE 2991 cell separator used in the purification of human pancreatic islets of Langerhans. Cryo Lett. 2010;31(4):310–7.
160. Chadwick DR, Robertson GS, Toomey P, Contractor H, Rose S, James RF, et al. Pancreatic islet purification using bovine serum albumin: the importance of density gradient temperature and osmolality. Cell Transplant. 1993;2(4):355–61.
161. Shimoda M, Itoh T, Iwahashi S, Takita M, Sugimoto K, Kanak MA, et al. An effective purification method using large bottles for human pancreatic islet isolation. Islets. 2012;4(6):398–404. doi:10.4161/isl.23008.
162. Shimoda M, Noguchi H, Fujita Y, Takita M, Ikemoto T, Chujo D, et al. Islet purification method using large bottles effectively achieves high islet yield from pig pancreas. Cell Transplant. 2012;21(2–3):501–8. doi:10.3727/096368911X605411.
163. Steensgaard J. Construction of isokinetic sucrose gradients for rate-zonal centrifugation. Eur J Biochem. 1970;16(1):66–70.
164. Lau D, Hering BJ, El-Ouaghlidi A, Jahr H, Brandhorst H, Brandhorst D, et al. Isokinetic gradient centrifugation prolongs survival of pig islets xenografted into mice. J Mol Med. 1999;77(1):175–7.
165. Freshney RI. Defined media and supplements. Culture of animal cells. New York: Wiley; 2010. p. 99–114.
166. Holmes MA, Clayton HA, Chadwick DR, Bell PR, London NJ, James RF. Functional studies of rat, porcine, and human pancreatic islets cultured in ten commercially available media. Transplantation. 1995;60(8):854–60.
167. Brandhorst H, Brandhorst D, Lau D, Hering BJ, Federlin K, Bretzel RG. Glucose sensitivity of porcine and human islets in vitro. J Mol Med. 1999;77(1):90–2.
168. Hering BJ, Kandaswamy R, Harmon JV, Ansite JD, Clemmings SM, Sakai T, et al. Transplantation of cultured islets from two-layer preserved pancreases in type 1 diabetes with anti-CD3 antibody. Am J Transplant. 2004;4(3):390–401.

169. Kerr-Conte J, Vandewalle B, Moerman E, Lukowiak B, Gmyr V, Arnalsteen L, et al. Upgrading pretransplant human islet culture technology requires human serum combined with media renewal. Transplantation. 2010;89(9):1154–60. doi:10.1097/TP.0b013e3181d154ac.
170. Avgoustiniatos ES, Scott WE, Suszynski TM, Schuurman HJ, Nelson RA, Rozak PR, et al. Supplements in human islet culture: human serum albumin is inferior to fetal bovine serum. Cell Transplant. 2012;21(12):2805–14. doi:10.3727/096368912X653138.
171. Noguchi H, Naziruddin B, Shimoda M, Fujita Y, Chujo D, Takita M, et al. Comparison of fresh and cultured islets from human and porcine pancreata. Transplant Proc. 2010;42(6):2084–6. doi:10.1016/j.transproceed.2010.05.098. doi:S0041-1345(10)00763-3 [pii].
172. Tai J, Tsang A, Tze WJ. Effects of serum and medium supplements, pH, and temperature on the viability of cultured porcine islets. Transplant Proc. 1994;26(2):818–9.
173. Sakamoto K, Hatakeyama E, Kenmochi T, Yamada K, Iwashita C, Asano T, et al. Improvement of porcine islet culture with porcine serum. Transplant Proc. 1998;30(2):391–2.
174. Brandhorst D, Brandhorst H, Hering BJ, Bretzel RG. Long-term survival, morphology and in vitro function of isolated pig islets under different culture conditions. Transplantation. 1999;67(12):1533–41.
175. Malaisse WJ, Sener A, Carpinelli AR, Anjaneyulu K, Lebrun P, Herchuelz A, et al. The stimulus-secretion coupling of glucose-induced insulin release. XLVI. Physiological role of L-glutamine as a fuel for pancreatic islets. Mol Cell Endocrinol. 1980;20(2):171–89. doi:0303-7207(80)90080-5 [pii].
176. Malaisse WJ, Sener A, Malaisse-Lagae F, Welsh M, Matthews DE, Bier DM, et al. The stimulus-secretion coupling of amino acid-induced insulin release. Metabolic response of pancreatic islets of L-glutamine and L-leucine. J Biol Chem. 1982;257(15):8731–7.
177. Malaisse WJ, Sener A, Malaisse-Lagae F, Hutton JC, Christophe J. The stimulus-secretion coupling of amino acid-induced insulin release. Metabolic interaction of L-glutamine and 2-ketoisocaproate in pancreatic islets. Biochim Biophys Acta. 1981;677(1):39–49.
178. Sener A, Malaisse WJ. Stimulation of insulin release by L-glutamine. Mol Cell Biochem. 1980;33(3):157–9.
179. Sener A, Malaisse-Lagae F, Malaisse WJ. The stimulus-secretion coupling of glucose-induced insulin release. Environmental influences on L-glutamine oxidation in pancreatic islets. Biochem J. 1982;202(2):309–16.
180. Khan K, Elia M. Factors affecting the stability of L-glutamine in solution. Clin Nutr. 1991;10(4):186–92. doi:0261-5614(91)90037-D [pii].
181. Heeneman S, Deutz NE, Buurman WA. The concentrations of glutamine and ammonia in commercially available cell culture media. J Immunol Methods. 1993;166(1):85–91.
182. Petronini PG, Urbani S, Alfieri R, Borghetti AF, Guidotti GG. Cell susceptibility to apoptosis by glutamine deprivation and rescue: survival and apoptotic death in cultured lymphoma-leukemia cell lines. J Cell Physiol. 1996;169(1):175–85. 10.1002/(SICI)1097-4652(199610)169:1 < 175::AID-JCP18 > 3.0.CO;2-C.
183. Brandhorst H, Duan Y, Iken M, Bretzel RG, Brandhorst D. Effect of stable glutamine compounds on porcine islet culture. Transplant Proc. 2005;37(8):3519–20.
184. Heldmaier G, Ruf T. Body temperature and metabolic rate during natural hypothermia in endotherms. J Comp Physiol B. 1992;162(8):696–706.
185. Escolar JC, Hoo-Paris R, Castex C, Sutter BC. Effect of low temperatures on glucose-induced insulin secretion and glucose metabolism in isolated pancreatic islets of the rat. J Endocrinol. 1990;125(1):45–51.
186. Ikemoto T, Noguchi H, Fujita Y, Takita M, Shimoda M, Sugimoto K, et al. New stepwise cooling system for short-term porcine islet preservation. Pancreas. 2010;39(7):960–3. doi:10.1097/MPA.0b013e3181dab19a.
187. Ilieva A, Yuan S, Wang R, Duguid WP, Rosenberg L. The structural integrity of the islet in vitro: the effect of incubation temperature. Pancreas. 1999;19(3):297–303.
188. Rijkelijkhuizen JK, van der Burg MP, Tons A, Terpstra OT, Bouwman E. Pretransplant culture selects for high-quality porcine islets. Pancreas. 2006;32(4):403–7.

Isolation of Pancreatic Islets from Nonhuman Primates

5

Dora M. Berman

Abstract

Nonhuman primates (NHP) constitute a highly relevant pre-clinical animal model to develop strategies for beta cell replacement. The close phylogenetic and immunologic relationship between NHP and humans results in cross-reactivity of various biological agents with NHP cells, as well as a very similar cytoarchitecture between islets from human and NHP that is strikingly different from that observed in rodent islets. The composition and location of endocrine cells in human or NHP islets, randomly distributed and associated with blood vessels, have functional consequences and a predisposition for paracrine interactions. Furthermore, translation of approaches that proved successful in rodent models to the clinic has been limited. Consequently, data collected from NHP studies can form the basis for an IND submission to the FDA. This chapter describes in detail the key aspects for isolation of islets from NHP, from organ procurement up to assessment of islet function, comparing and emphasizing the similarities between isolation procedures for human and NHP islets.

Keywords

Islet isolation • Islet purification • Islet yield • Islet assessment

5.1 Introduction

Several centers have reported extended pancreatic islet allograft survival with restoration of glycemic control and insulin independence in patients with type 1 diabetes [1–3]. This relatively minimally invasive procedure to replace insulin producing cells has proved to improve quality of life and halt progression of complications in diabetic patients [4, 5]. Nevertheless,

D.M. Berman (✉)
Diabetes Research Institute, University of Miami
Leonard M. Miller School of Medicine,
1450 NW 10 Avenue, Miami, FL, USA
e-mail: dberman2@med.miami.edu

M. Ramírez-Domínguez (ed.), *Pancreatic Islet Isolation*, Advances in Experimental Medicine and Biology 938, DOI 10.1007/978-3-319-39824-2_5

widespread application of this technique still faces important challenges, such as progressive graft dysfunction, need for chronic immune suppression, need for multiple donors and allo sensitization, among others [6]. Nonhuman primates (NHP) constitute a highly relevant pre-clinical animal model that should allow for rapid, direct translation of experimental results to clinical trials. The animal species used in submission of an IND for the FDA should mimic as close as possible the profile found in humans. In this regard, the NHP model to develop strategies for beta cell replacement fulfills several of the required characteristics: the close phylogenetic and immunologic relationship between NHP and humans results in cross-reactivity of various biological agents with NHP cells. The insulin content of NHP islets is similar to that in human islets, and although the insulin secretory response to glucose is higher in NHP islets [7], the cytoarchitecture of human and NHP islets is very similar, and strikingly different from that observed in rodent islets [8]. The composition and position of endocrine cells in human or NHP islets, randomly distributed and associated with blood vessels, have functional consequences and a predisposition for paracrine interactions [8]. Additionally, the fact that NHP are upright imposes physical constraints on graft sites similar to those in humans. Brain death in cadaveric donors is characterized by activation of proinflammatory cytokines that affect the pancreas and islet function [9, 10]. Therefore, access to NHP islets from healthy donors provides a useful model to study islet cell biology and function. Furthermore, translation of approaches that proved successful in rodent models to the clinic has been limited. Consequently, data collected from NHP studies can form the basis for an IND submission to the FDA. Human islets used for clinical purposes are manufactured under current good-manufacturing-practice regulatory guidelines. This involves a costly process to establish an infrastructure in place to address all issues for regulatory compliance. Laboratories doing NHP islet isolation should try to adhere to Good Laboratory Practices (GLP) as much as possible, so that any promising finding in this valuable model can be included in IND submissions to FDA and be translated to the clinic. NHP islet isolation involves a mechanically enhanced enzymatic digestion, following the semi-automatic procedure used to isolate human islets [11–14] with minor modifications. The weight of the pancreas in NHP is smaller compared to that in humans. For example, the weight of the pancreas in 246 male cynomolgus monkeys weighing 6.8 ± 1.7 kg with an age of 6.6 ± 2.0 years (n = 238 for the age variable) was 10.8 ± 2.3 g (mean ± SD, unpublished data) which is approximately ten times smaller than the weight reported for human pancreas, i.e., Schaefer J.H. reported 84.88 ± 14.95 g in females and 90.3 ± 15.08 g in males in 216 cases [15]; recently, Caglar et al. [16] reported a pancreas weight of 87.3 ± 30.6 g in 114 males between 25 and 88 years old. Therefore, the Ricordi chamber we mostly use for NHP studies has a volume of 250 ml vs the 500 ml used for human pancreas processing. Nevertheless, occasionally, more than one organ may be processed together. In general, for NHP isolations where the weight of pancreas used is < 30 g we use the 250 ml chamber, and for a total pancreas weight >30 g we use the 500 ml Ricordi chamber.

5.2 Description of the Procedure and Key Aspects

5.2.1 Laboratory Set Up

We conduct our islet isolation following the Standard Operation Procedures (SOP) that we wrote aiming to perform the process the closest possible to the standards of a GLP laboratory. An SOP is a document that describes in detail processes to be performed at an institution, and they should enable any trained individual to execute the procedure successfully. By following SOPs the institution guaranties quality, reproducibility, traceability and integrity of information and data. All instruments used for the NHP islet isolation have to be serviced annually or semi-annually, as suggested by the manufacturers, and all the documentation has to be kept for the period of time required by the institution. This includes micropipettes, microscopes, biosafety cabinets, centrifuges, incubators, refrigerators, freezers, etc. Sterile conditions must be maintained throughout the procedure.

5.2.1.1 Procedures Prior to the Day of Islet Isolation

Several solutions can be made few days before the islet isolation, and all serum to be used in the solutions must be heat inactivated at 56 °C for 30 min to destroy the complement present. Solutions made in the lab or purchased that are not sterile must be filter sterilized (0.22 µm filter). Addition of nicotinamide to media used in islet isolation has been shown to improve islet yields [17], and stock solutions (2.5 M in DPBS) of nicotinamide can be frozen at −20 °C and be added to the corresponding media on the day of islet isolation. Examples of solutions prepared in advanced include dithizone to stain islets [18], Hanks with 2 % fetal bovine serum (FBS) to recirculate in the Ricordi chamber, washing solution (RPMI with 10 % FBS; 1 M Hepes and 1 M NaOH), culture media (CMRL with 10 % FBS) and Eurocollins (supplemented the day before isolation with Electrolyte additive solution). One or two bags of sterile saline are placed at −20 °C. Additionally, all instruments and devices that will be used for the islet isolation, e.g., tray with scissors, forceps, etc.; Ricordi chamber with attached Masterflextubings, etc. can be washed and autoclaved in advanced.

5.2.1.2 Procedures on the Day of Islet Isolation

Solutions and Instruments

On the day of isolation, Nicotinamide is added to Eurocollins solution (5 mM) that will be used to transport the organ/s to the laboratory and to culture medium (10 mM); washing solution is finalized by adding insulin (Humulin®R, 20 U/L); Heparin (10,000 U/L) and Nicotinamide (10 mM). Following sterile technique, the autoclaved package containing isolation instruments, Ricordi chamber with five silicon nitride marbles inside and Masterflex tubings is placed in a biosafety cabinet, opened and the material is organized. A temperature probe is connected to the chamber, the chamber is closed and the ends of the tubing connected to the bottom of the chamber (size 16) and to the top of the chamber (size 17) are placed inside an empty 1 L beaker. Part of the size 16 tube, which contains a metal solenoid, is taken out from the safety cabinet and is connected to a peristaltic pump that will push the liquid from the 1 L beaker, through the solenoid into the chamber. Once the chamber is filled, the liquid will be pushed out via the size 17 tube, which will drain the liquid into the beaker, creating a closed circuit (Fig. 5.1). Hanks with 2 % FBS is added to the beaker to fill the closed circuit, with the purpose of verifying there are no leaks in the circuit, and the serum present in the solution will preclude the enzyme solution that will be used later on to stick to the tubings and walls of the chamber. After this test, the chamber is emptied and ready to be filled with tissue and digestive enzymes.

Preparation of Digestive Enzymes

The cocktail of digestive enzymes used is the same as the one used for human pancreas, scaling down the amounts to adjust to the size of the NHP pancreas. We used Liberase HI (Roche Diagnostics, Indianapolis, Indiana), a specialized blend of collagenase and proteases enzymes until its use in human islet transplantation stopped because of potential risk of bovine spongiform

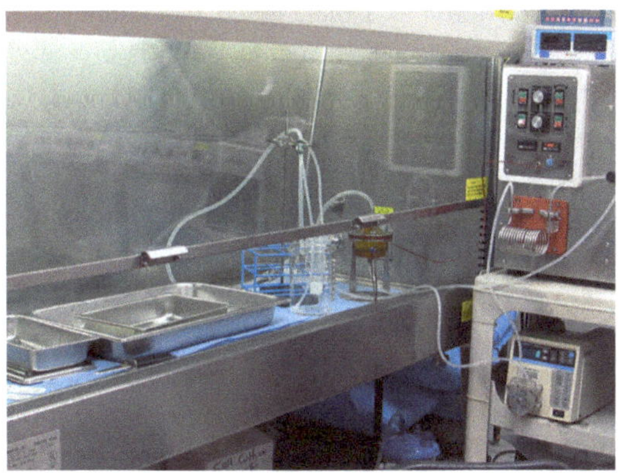

Fig. 5.1 Set up for pancreas processing in a biosafety cabinet including connection of tubings to a peristaltic pump located outside the cabinet (in the *lower* shelf of a transporting cart)

encephalopathy in this enzyme blend [19]. Since 2009, we have been using mammalian tissue free Liberase (Liberase MTF C/T, Roche Diagnostics, Indianapolis, Indiana), where collagenases and thermolysin are prepared separately and mixed just before the organ digestion. In the lots of Liberase MTF C/T we have used so far, each bottle of lyophilized collagenases I and II contains approximately 2800–3100 U (Wünsch, calculated), and each bottle of lyophilized thermolysin contains approximately 150,000–215,000 Units. Currently, we are using 33 % of a collagenase vial and 40 % of a thermolysin vial when processing organs in a 250 ml Ricordi Chamber. However, these proportions may need to be readjusted for different lots as well as for a larger chamber. The final preparation of collagenase in Hanks contains $CaCl_2$ (2.7 mM), HEPES (21 mM) and NaOH (2.5 mM) to achieve a final pH of 7.4. Thermolysin diluted in Hanks is prepared in a separate container and both enzymes are kept at 4 °C.

Pancreas Procurement

All procedures performed must be previously approved by the institutional animal care and use committee. The best opportunity for healthy islets is derived from donors >4 years old and weighing 4 kg or more. Donor animals are NPO and placed under general endotracheal anesthesia. After appropriate prepping and draping, a midline incision is performed (xyphoid to pubis), the distal aorta is isolated and cannulated (IV infusion set), and the proximal aorta is prepared for cross-clamping (cephalad to superior mesenteric artery). Subsequent to exsanguination and collection of appropriate blood samples, gentle dissection of the pancreas is initiated with exposure of portal vein and splenic vessels. The pancreas is then mobilized with marking suture ligatures on the main and accessory pancreatic ducts. The pancreatic head is freed anterior and posterior from the portal vein and mobilization from spleen, portal vein and duodenum are then completed. The excised pancreas is placed in cold Eurocollins solution. Additional required tissue is then removed (e.g., lymph nodes, etc.). A pancreas that requires delay or transportation is perfused (gravity only) with cold UW solution prior and during resection (cephalad to the aorta clamped). Topical ice may also be used in lesser sac and an inferior aortic catheter is used for perfusion. Survivor donors are not perfused in conjunction with meticulous hemostasis. For shipping, the pancreas is transferred to a Nalgene container with cold UW solution covering the organ and the container is placed in one to two sterile isolation bags and packed into ice. Prompt shipping is then initiated.

Collagenase Digestion and Dilution

Figure 5.1 shows the set up ready in the lab before the arrival of the pancreas. Part of the set up in the biosafety cabinet includes two trays, a smaller one where the pancreas will be cannulated and distended, laying inside a larger pan. Once the organ arrives to the lab, and while it is being weighed, a bag of frozen saline is placed in the larger pan, chipped using an ice pick, and suspended in 1 L cold saline. The smaller tray inside the big pan is filled with cold Eurocollins solution. This way, if the pancreas falls off the inside tray, it will fall into saline and not into water. The objective of ductal distension with collagenase is to deliver the enzyme to the connective tissue and dissociate the acinar cells surrounding the islets, thereby releasing intact islets. In order to maximally distend the organ with the enzyme, it is cut in the middle (body of the organ) using a sharp movement with a scalpel. The pancreatic duct in each half is visualized and cannulated with a small catheter (i.e., 26 G or larger diameter, depending on the size of the duct) which is then secured with silk suture (Fig. 5.2). After cannulation, the solution in the small tray, as well as the larger pan are removed, and the distension of the gland is done are room temperature. In the meantime, the preparation of collagenase is being warmed up in a water bath at 48 °C until it reaches room temperature (23–24 °C). Immediately before organ distension, thermolysin is added to the preparation of collagenase, as well as one vial of DNAse (Pulmozyme®) to prevent cell clumping. The cocktail of enzymes is poured into a 500 ml beaker and each half of the pancreas is distended manually two times using 30 ml

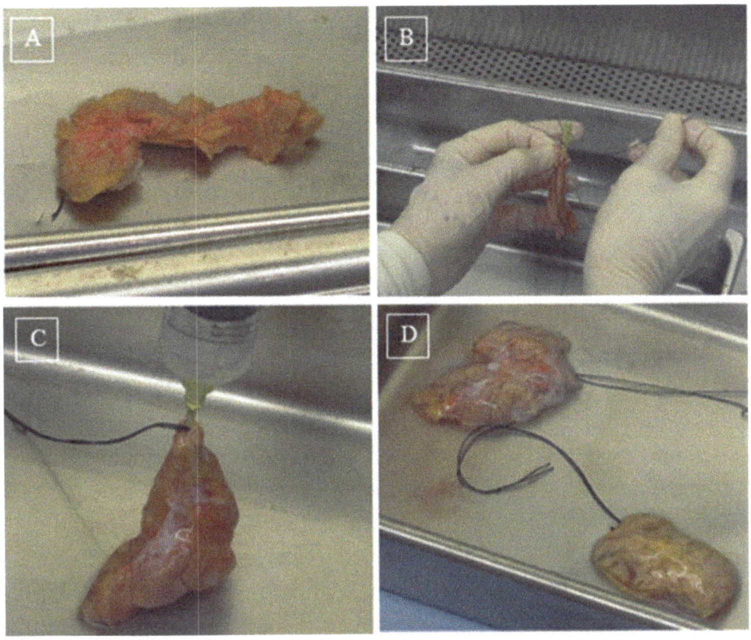

Fig. 5.2 Pancreas cannulation and distension. (**a**) NHP pancreas; (**b**) cannulation of one half of the organ; (**c**) distension with collagenase/thermolysin mixture; and (**d**) two halves of the organ distended with the enzyme

syringes filled with enzyme cocktail pushing at a slow, constant pace (Fig. 5.2). Once ductal distension is complete, the two pieces of pancreas are trimmed and placed into the lower half of the Ricordi chamber together with the rest of the enzyme contained in the tray and 5 silicon nitride marbles that will help the digestion of the organ mechanically. A stainless steel screen (533 μm mesh) is placed on top of the tissue to retain undigested tissue and the chamber is closed. The tubing feeding the chamber (size 16) and the tubing collecting the effluent from the chamber (size 17) are both placed into a 250 ml conical and the leftover enzyme cocktail that was not used for distention is poured into the conical. The perfusion pump is turned on at a flow rate of 200 ml/min and the solenoid is placed into a water bath (at 48 °C) to warm up the contents of the chamber until it reaches 37 °C. Once the closed circuit has been filled, the level of the enzyme cocktail is kept at 75 ml in the conical, so that the total volume of liquid recirculating the chamber is approximately 400 ml. Warm (37 °C) Hanks solution is added if necessary to achieve the desired volume for recirculation. At this moment, the flow rate is lowered to 150 ml/min, a timer is turned on, the chamber is brought outside the biosafety cabinet and manual, gentle shaking begins and will continue for the first 20 min of the process. Afterwards, the chamber may be placed in an automated shaker until the end of the dilution. Initially, the shaking is done using very delicate movements, until the temperature inside the chamber reaches 37 °C. The temperature in the chamber is kept at 37 °C by moving the solenoid in and out of the 48 °C water bath. Once the tissue begins to dissociate, samples from the chamber (coming from the size 17 tubing) are taken and placed directly into small dishes containing dithizone, to determine the presence of isolated islets (stained in red), as well as the breakup of acinar tissue under the microscope at 40× magnification. The determination of the time to stop the digestion process is critical, as stopping too early results in lots of islets still embedded in acinar tissue, while stopping too late results in partially digested islets. The length of time for digestion depends on the tissue, on the ratio of thermolysin/collagenase used, and often times on the lot of enzyme used. Once the determination to stop digestion has been made, which can be within a range of approximately 4–10 min of recirculation of the enzyme in the case of NHP pancreas, 100 ml of FBS are added to the closed circuit to inactivate the enzyme, the solenoid is simultaneously placed in an iced water bath and the flow rate is raised to 200 ml/min. From that moment on, all solutions used are cold and kept on ice. The circuit is then opened, by placing the size 16 tubing into a 2 L flask containing 2 L washing solution and the effluent from the chamber (size 17 tubing) in a 4 L flask containing 1 L washing solution (Fig. 5.3). The contents of the 250 ml conical containing digested tissue are poured into the 4 L flask, together with the contents obtained after washing the conical two times with washing solution. Once 3 L are collected into the 4 L flask, it is removed and the rest of the media washing the chamber is collected in 1 L bottles. This process continues until no more free islets are detected in samples taken from the chamber. Ideally, only duct tissue and blood vessels should remain in the chamber, and the weight of the remaining tissue must be subtracted from the weight of tissue originally placed in the chamber.

Concentration and Purification of Islets

The dissociated tissue is collected in 250 ml conicals kept on ice, which are spun in refrigerated centrifuges for 1 min at 1000 rpm (Fig. 5.4a). After discarding the supernatants, pellets are pooled and the process is repeated until all the dissociated tissue is collected into one conical tube. The next step is islet purification. In general, we have less than 10 ml of digested tissue, and proceed to purify islets using a discontinuous gradient with Ficoll (polysucrose, Euro-Collins base, Mediatech), in a similar way as the one described for human islets [11, 20, 21]. Briefly, a sterile closed system is provided by using the COBE 2991 cell processor (COBE Laboratories, Inc., Lakewood, Colorado, USA), and ideally this procedure should to be done in a refrigerated cell processing room at 4 °C. The digested tissue is resuspended in 300 ml stock Ficoll (density 1.132 g/mL), loaded into a 600 ml transfer bag (Fig. 5.4b) and it is bottom loaded by gravity into the doughnut-shaped COBE bag (Fig. 5.4c). The discontinuous gradient is obtained by applying subsequently Ficoll solutions with density 1.108, 1.096 and 1.037 g/mL (75 mL each) and 50 ml Hanks at 2400 rpm. After a 3-min centrifugation, four fractions are collected into each of 250 ml conicals containing 100 ml of 10 % RPMI kept on ice: a 85 mL (layer #1), and 75 ml into each of the subsequent conicals (layers 2–4). The purest islets are generally found in layer #2, at the interface of 1.037/1.096 densities; less pure islets are found at the interface of 1.096/1.108 densities (layer 3); and the least pure islets are in layer 4, at the interface of 1.108/1.132 gradients. A sample from the tissue remaining in the COBE bag is taken to determine the presence of isolated islets or embedded islets after staining with dithizone. The conicals containing the islets are filled with

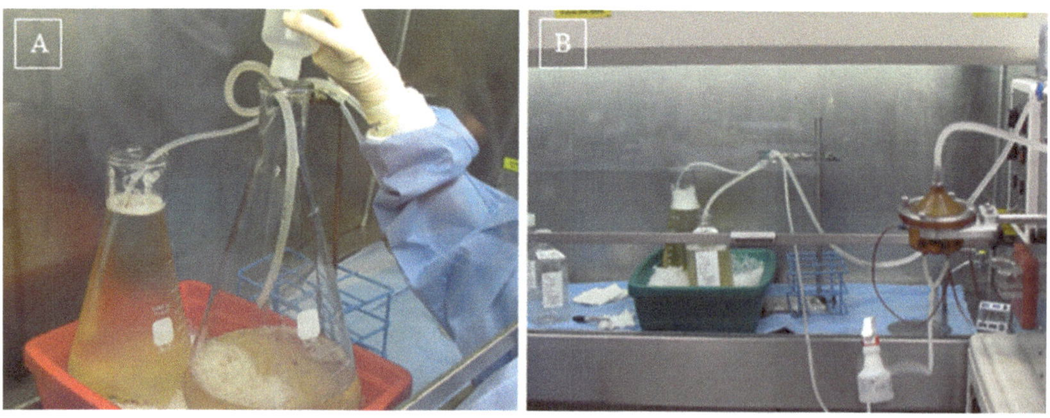

Fig. 5.3 Collection of digested organ. (**a**) After stopping enzymatic activity, collected digested tissue is poured into a flask containing 1 L washing media; (**b**) Perfusing the chamber with washing media and collecting effluent from the chamber into 1 L bottles

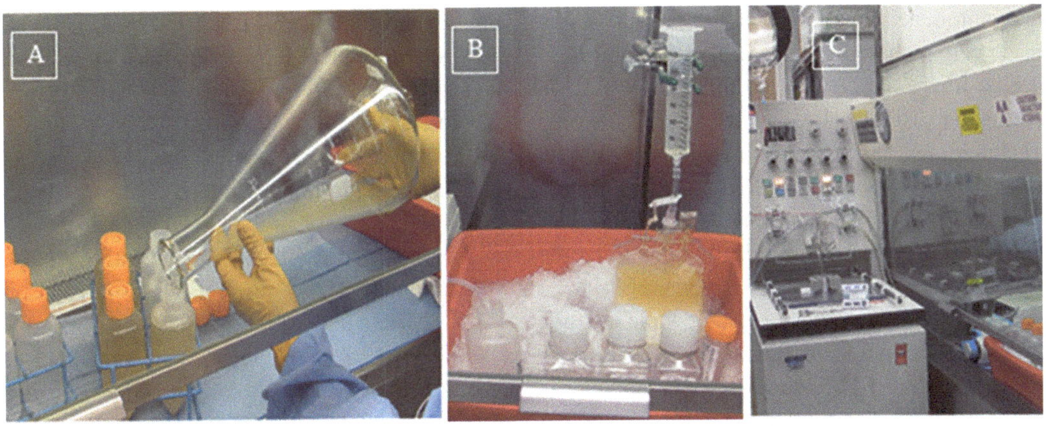

Fig. 5.4 Concentration and purification of islets. (**a**) Digested tissue is collected into 250 ml conicals; (**b**) concentrated digest resuspended in stock Ficoll is poured into a 600 ml transfer bag; (**c**) contents of the transfer bag are emptied into the doughnut-shaped COBE bag located into the cell processor

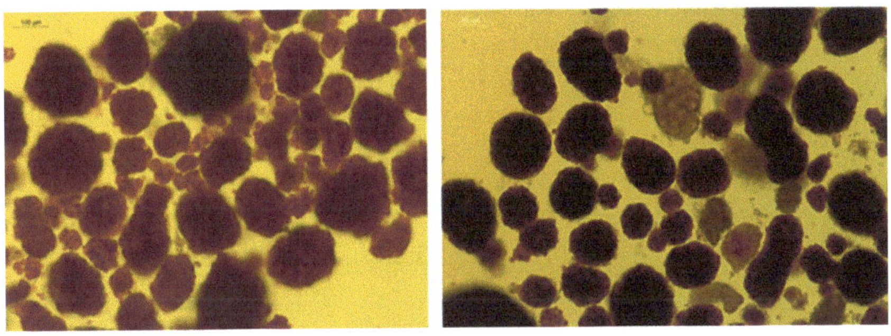

Fig. 5.5 Representative samples of purified NHP islets. Purified isolated islets from two different pancreata stained with dithizone. Bar in the picture indicates 100 μm. Magnification for both pictures at 100×

10 % RPMI and centrifuged at 1,500 rpm for 3 min at 4 °C. After removing the supernatant from each of the conicals, each pellet is resuspended in a final volume of 100 ml 10 % RPMI for layers 1 and 4 and 100 ml culture media at room temperature for layers 2 and 3. Figure 5.5 shows representative pictures of purified NHP islets stained with dithizone. They are shaped similar to human islets, with heterogeneous shapes and sizes. In general, islets obtained from the purest layer after purification are >90 % pure.

Determination of Islet Yield and Purity

Samples from each conical are used to count and assess the islet yield. We generally perform a 1:500 dilution, by taking 0.2 ml from the 100 ml suspension and placing the aliquot into a counting dish with dithizone and Hanks. Samples are counted using a microscope with a calibrated eye piece at a 40× magnification. Using 50 μm diameter increments, islets are divided into seven classes, as previously described [22]: 50–100; 100–150; 150–200; 200–250; 250–300; 300–350 and >350. Calculating the mean volume for each diameter class and relative conversion into equivalent number of islets with a diameter of 150 μm gives the yield in islet equivalents (IEQ) [22]. There is an inherent variability in the counting of islet numbers between operators. In fact, it is rare that two operators counting the same islet sample would attain the same result, and it has been estimated that standard manual methods can give intra- and inter-operator variabilities (CV) of >20 % [23]. To circumvent this issue, there are now automated counters to quantitatively assess islet cell numbers using fully computerized digital image analysis-based methods. Using an automatic islet cell counter (ICC; Biorep Technologies) that uses a digital imager and an image analysis segmentation method implemented in LabVIEW, we obtained good correlation (r^2>0.95) between the IEQ obtained by the ICC and the same experienced operator [24]. One limitation of the ICC we use is that while it works well with very pure islet preparations, as the ones we obtain from NHP pancreas, it sometimes has problems with less pure preparations including those from human pancreata. Nevertheless, efforts are ongoing to standardize updated versions of ICC that are able to distinguish endocrine from exocrine tissue and will be ideal to use with islet preparations obtained from human pancreas. The purity of the preparation is still estimated by comparing the relative proportions of dithizone stained tissue (red) vs dithizone negative tissue (lighter colored exocrine tissue).

Islet Assessment in NHP Islets

In Vitro Tests

The fundament of the in vitro tests used to assess islet function is based on the measurement of insulin release after stimulation with different glucose concentrations. In this regard, glucose challenge can be performed in static incubations or during continued, dynamic perifusion of islets, and a stimulation index can be calculated as the ratio between stimulated and basal insulin release [25, 26]. Analysis of predictors of successful human islet transplant from the collaborative Islet Transplant Registry (CITR) from 1999 to 2010 showed no association between the stimulation index from static incubations and clinical outcome [27]. We routinely asses our islet preparations using a perifusion assay performed 48 h after islet isolation as previously described [25]. This assay assesses not only the release of preformed insulin, but also the one newly synthetized. We found a positive correlation between the stimulation index obtained from perifusion studies and the earliest fasting c-peptide levels measured on post-operative day (POD) three to five in recipients of a marginal mass of allogeneic islets under the cover of steroid-free immune suppression [28].

In Vivo Test

Assessment of islet function in vivo involves transplantation of human or NHP islets into nude (athymic) mice. Because of a congenital thymic aplasia, nude mice do not reject transplanted xenogenic islets [29, 30]. Transplantation of viable NHP islets under the kidney capsule of streptozotocin (STZ)-induced diabetic nude mice results in diabetes reversal, with stable non-fasting blood glucose values. Restoration of hyperglycemia after nephrectomy of the transplanted kidney confirms the reversal of diabetes was due to the transplanted islets and not to residual function of the native pancreas [26]. In general, transplantation of 2,000 IEQ from NHP results in diabetes reversal in this system [26]. However, no correlation has been established between reversal of diabetes in immunodeficient mice and NHP graft outcome in islet allotransplantation.

5.3 Conclusions

The close phylogenetic and immunologic relationship between NHP and humans makes NHP a highly relevant pre-clinical animal model that should allow for rapid, direct translation of experimental results in transplantation of insulin producing cells to clinical trials. Consequently, data collected from NHP studies can form the basis for an IND submission to the FDA. The costly process involved in the isolation, purification and functional assessment of NHP islets is very similar to that used to obtain human islets, with some nuances emphasized in this chapter. Similar to other species, NHP islet isolation is a craftsmanship where the experience of personnel involved in the process plays a critical role. Nevertheless, as we continue streamlining the process, we hope to get closer to a standardization and optimization of the procedure.

Acknowledgement I want to thank Dr. Norma S. Kenyon, whose expert guidance and support enabled me to become proficient in the field of NHP islet isolation.

References

1. Barton FB, Rickels MR, Alejandro R, Hering BJ, Wease S, Naziruddin B, Oberholzer J, Odorico JS, Garfinkel MR, Levy M, Pattou F, Berney T, Secchi A, Messinger S, Senior PA, Maffi P, Posselt A, Stock PG, Kaufman DB, Luo X, Kandeel F, Cagliero E, Turgeon NA, Witkowski P, Naji A, O'Connell PJ, Greenbaum C, Kudva YC, Brayman KL, Aull MJ, Larsen C, Kay TW, Fernandez LA, Vantyghem MC, Bellin M. Shapiro AM: improvement in outcomes of clinical islet transplantation: 1999–2010. Diabetes Care. 2012;35:1436–45.
2. Froud T, Ricordi C, Baidal DA, Hafiz MM, Ponte G, Cure P, Pileggi A, Poggioli R, Ichii H, Khan A, Ferreira JV, Pugliese A, Esquenazi VV, Kenyon NS, Alejandro R. Islet transplantation in type 1 diabetes mellitus using cultured islets and steroid-free immunosuppression: Miami experience. Am J Transplant. 2005;5:2037–46.
3. Harlan DM, Kenyon NS, Korsgren O, Roep BO. Current advances and travails in islet transplantation. Diabetes. 2009;58:2175–84.
4. Cure P, Pileggi A, Froud T, Messinger S, Faradji RN, Baidal DA, Cardani R, Curry A, Poggioli R, Pugliese A, Betancourt A, Esquenazi V, Ciancio G, Selvaggi G, Burke 3rd GW, Ricordi C, Alejandro R. Improved metabolic control and quality of life in seven patients with type 1 diabetes following islet after kidney transplantation. Transplantation. 2008;85:801–12.
5. Bassi R, Fiorina P. Impact of islet transplantation on diabetes complications and quality of life. Curr Diab Rep. 2011;11:355–63.
6. Mineo D, Pileggi A, Alejandro R, Ricordi C. Point: steady progress and current challenges in clinical islet transplantation. Diabetes Care. 2009;32:1563–9.
7. Mueller KR, Balamurugan AN, Cline GW, Pongratz RL, Hooper RL, Weegman BP, Kitzmann JP, Taylor MJ, Graham ML, Schuurman HJ, Papas KK. Differences in glucose-stimulated insulin secretion in vitro of islets from human, nonhuman primate, and porcine origin. Xenotransplantation. 2013;20:75–81.
8. Cabrera O, Berman DM, Kenyon NS, Ricordi C, Berggren PO, Caicedo A. The unique cytoarchitecture of human pancreatic islets has implications for islet cell function. Proc Natl Acad Sci U S A. 2006;103:2334–9.
9. Contreras JL, Eckstein C, Smyth CA, Sellers MT, Vilatoba M, Bilbao G, Rahemtulla FG, Young CJ, Thompson JA, Chaudry IH, Eckhoff DE. Brain death significantly reduces isolated pancreatic islet yields and functionality in vitro and in vivo after transplantation in rats. Diabetes. 2003;52:2935–42.
10. Rech TH, Crispim D, Rheinheimer J, Barkan SS, Osvaldt AB, Grezzana Filho TJ, Kruel CR, Martini J, Gross JL, Leitao CB. Brain death-induced inflammatory activity in human pancreatic tissue: a case-control study. Transplantation. 2014;97:212–9.
11. Ricordi C, Lacy PE, Finke EH, Olack BJ, Scharp DW. Automated method for isolation of human pancreatic islets. Diabetes. 1988;37:413–20.
12. Berman DM, O'Neil JJ, Coffey LC, Chaffanjon PC, Kenyon NM, Ruiz Jr P, Pileggi A, Ricordi C, Kenyon NS. Long-term survival of nonhuman primate islets implanted in an omental pouch on a biodegradable scaffold. Am J Transplant. 2009;9: 91–104.
13. Kenyon NS, Chatzipetrou M, Masetti M, Ranuncoli A, Oliveira M, Wagner JL, Kirk AD, Harlan DM, Burkly LC, Ricordi C. Long-term survival and function of intrahepatic islet allografts in rhesus monkeys treated with humanized anti-CD154. Proc Natl Acad Sci U S A. 1999;96:8132–7.
14. Kenyon NS, Fernandez LA, Lehmann R, Masetti M, Ranuncoli A, Chatzipetrou M, Iaria G, Han D, Wagner JL, Ruiz P, Berho M, Inverardi L, Alejandro R, Mintz DH, Kirk AD, Harlan DM, Burkly LC, Ricordi C. Long-term survival and function of intrahepatic islet allografts in baboons treated with humanized anti-CD154. Diabetes. 1999;48:1473–81.
15. Schaefer JH. The normal weight of the pancreas in the adult human being: a biometric study. Anat Rec. 1926;32:119–32.
16. Caglar V, Kumral B, Uygur R, Alkoc OA, Ozen OA, Demirel H. Study of volume, weight and size of normal pancreas, spleen and kidney in adults autopsies. Forensic Med Anat Res. 2014;2:63–9.
17. Ichii H, Wang X, Messinger S, Alvarez A, Fraker C, Khan A, Kuroda Y, Inverardi L, Goss JA, Alejandro R, Ricordi C. Improved human islet isolation using nicotinamide. Am J Transplant. 2006;6:2060–8.
18. Latif ZA, Noel J, Alejandro R. A simple method of staining fresh and cultured islets. Transplantation. 1988;45:827–30.
19. Kenmochi T, Miyamoto M, Mullen Y. Protection of mouse islet isografts from nonspecific inflammatory damage by recipient treatment with nicotinamide and 15-deoxyspergualin. Cell Transplant. 1996;5:41–7.
20. Ichii H, Pileggi A, Molano RD, Baidal DA, Khan A, Kuroda Y, Inverardi L, Goss JA, Alejandro R, Ricordi C. Rescue purification maximizes the use of human islet preparations for transplantation. Am J Transplant. 2005;5:21–30.
21. London NJM, Robertson GSM, Chadwick DR, James RFL, Bell PRF. Adult islet purification. Austin: R.G. Lanades Company; 1995.
22. Ricordi C, Gray DW, Hering BJ, Kaufman DB, Warnock GL, Kneteman NM, Lake SP, London NJ, Socci C, Alejandro R, et al. Islet isolation assessment in man and large animals. Acta Diabetol Lat. 1990;27:185–95.
23. Kissler HJ, Niland JC, Olack B, Ricordi C, Hering BJ, Naji A, Kandeel F, Oberholzer J, Fernandez L, Contreras J, Stiller T, Sowinski J, Kaufman DB. Validation of methodologies for quantifying isolated human islets: an islet cell resources study. Clin Transplant. 2010;24:236–42.

24. Buchwald P, Berman DM, Khan A, Szust J, Rabassa A, Fraker C, Echeverri F, Bernal A, Inverardi L, Kenyon NS, Ricordi C. Assessment of the accuracy and reproducibility of a digital image analysis-based automatic islet cell counter (ICC) using microsphere mixtures as well as human and NHP islets. In: Cell Transplant Society – International Xenotransplantation Association (CTS-IXA) 2011 Joint Congress. Miami, Cell Medicine, p. P111.
25. Idahl LA. A micro perifusion device for pancreatic islets allowing concomitant recordings of intermediate metabolites and insulin release. Anal Biochem. 1972;50:386–98.
26. Ranuncoli A, Cautero N, Ricordi C, Masetti M, Molano RD, Inverardi L, Alejandro R, Kenyon NS. Islet cell transplantation: in vivo and in vitro functional assessment of nonhuman primate pancreatic islets. Cell Transplant. 2000;9:409–14.
27. Balamurugan AN, Naziruddin B, Lockridge A, Tiwari M, Loganathan G, Takita M, Matsumoto S, Papas K, Trieger M, Rainis H, Kin T, Kay TW, Wease S, Messinger S, Ricordi C, Alejandro R, Markmann J, Kerr-Conti J, Rickels MR, Liu C, Zhang X, Witkowski P, Posselt A, Maffi P, Secchi A, Berney T, O'Connell PJ, Hering BJ, Barton FB. Islet product characteristics and factors related to successful human islet transplantation from the Collaborative Islet Transplant Registry (CITR) 1999–2010. Am J Transplant. 2014;14:2595–606.
28. Berman DM, Cabrera O, Kenyon NM, Miller J, Tam SH, Khandekar VS, Picha KM, Soderman AR, Jordan RE, Bugelski PJ, Horninger D, Lark M, Davis JE, Alejandro R, Berggren PO, Zimmerman M, O'Neil JJ, Ricordi C, Kenyon NS. Interference with tissue factor prolongs intrahepatic islet allograft survival in a nonhuman primate marginal mass model. Transplantation. 2007;84:308–15.
29. Ricordi C, Scharp DW, Lacy PE. Reversal of diabetes in nude mice after transplantation of fresh and 7-day-cultured (24 degrees C) human pancreatic islets. Transplantation. 1988;45:994–6.
30. Ricordi C, Zeng YJ, Alejandro R, Tzakis A, Venkataramanan R, Fung J, Bereiter D, Mintz DH, Starzl TE. In vivo effect of FK506 on human pancreatic islets. Transplantation. 1991;52:519–22.

Necessities for a Clinical Islet Program

6

Wayne J. Hawthorne

Abstract

For more than two decades we have been refining advances in islet cell transplantation as a clinical therapy for patients suffering from type 1 diabetes. A great deal of effort has gone to making this a viable therapy for a broader range of patients with type 1 diabetes. Clinical results have progressively improved, demonstrating clinical outcomes *on par* with other organ transplants, specifically in terms of insulin independence, graft and patient survival. We are now at the point where islet cell transplantation, in the form of allotransplantation, has become accepted as a clinical therapy in adult patients affected by type 1 diabetes, in particular those suffering from severe hypoglycaemic unawareness. This chapter provides an overview on how this has been undertaken over the years to provide outcomes on par with other organ transplantation results. In particular this chapter focuses on the processes and facilities that are required to establish a clinical islet isolation and transplantation program. It also outlines the very important underpinning processes of selection of the organ donor for islet isolation, the processes of organ donor operation and preservation of the pancreas by various means and the ideal ways to best improve outcomes for human islet cell isolation. Providing these more optimal conditions we can underpin the isolation processes to provide islets for transplantation and as such a safe, effective and feasible therapeutic option for an increasing number of patients suffering from type 1 diabetes with severe hypoglycaemic unawareness.

W.J. Hawthorne (✉)
National Pancreas and Islet Transplant Laboratories,
The Westmead Institute for Medical Research,
Westmead, NSW 2145, Australia

Department of Surgery, Westmead Clinical School,
Westmead Hospital, University of Sydney, Westmead,
NSW 2145, Australia
e-mail: wayne.hawthorne@sydney.edu.au;
http://www.westmeadinstitute.org.au

M. Ramírez-Domínguez (ed.), *Pancreatic Islet Isolation*, Advances in Experimental Medicine and Biology 938, DOI 10.1007/978-3-319-39824-2_6

Keywords
Diabetes • Insulin • Islet • Islet cell • Islet cell allotransplantation • Islet cell autotransplantation • Islet cell isolation • Islet equivalent (IEQ), type 1 diabetes (T1D)

Abbreviations

BMI	Body mass index
CIT	Cold ischaemic time
CMRL	Connaught Medical Research Laboratories
CS	Celsior
HTK	Histidine-tryptophan-ketoglutarate
IEQ	Islet cell isolation, islet equivalent
IEQ/g	Islet equivalents per gram
MTC	Mixed treatment comparison
QALY	Quality-adjusted life years
SPK	Simultaneous pancreas and kidney
T1D	Type 1 diabetes
UW	University of Wisconsin solution

6.1 Introduction

For more than two decades we have been refining advances in islet cell transplantation as a clinical therapy for patients suffering from type 1 diabetes (T1D). A great deal of effort has gone to making this a viable therapy for a broader range of patients with T1D and clinical results have progressively improved, demonstrating clinical outcomes *on par* with other organ transplants, specifically in terms of insulin independence, graft and patient survival [1]. We are now at the point where islet cell transplantation, in the form of allotransplantation, has now become accepted as a clinical therapy in adult patients affected by T1D, in particular those suffering from severe hypoglycaemic unawareness.

Islet cell transplantation has also gained greater acceptance as a viable therapeutic option after pancreatectomy for painful chronic pancreatitis in the form of autotransplantation. Islet cell autotransplantation therapy has become widely accepted for this subpopulation of patients, seeing broader acceptance and earlier intervention to provide pain relief to these patients [2]. In this chapter we will however, be focusing directly on allotransplantation rather than autotransplantation, but it should be noted that there are a significant number of processes that are identical to both forms in the isolation and preparation of islets for transplantation. The clear overlaps between both types of transplant will become obvious to the reader and as such the way they are performed can be utilised in either process.

Overall we have seen significant improvements to isolation and transplantation results due to the significant research undertaken to improve outcomes. We have undertaken studies that have provided significant improvements to how we choose the type of donor pancreas, how we isolate the islets, how we culture them and ultimately ensure the islet preparation is suitable for transplantation [3]. On the recipient side we have further improved outcomes with changes to the transplant and to the pharmacological treatment of recipients. Anti-inflammatory treatments facilitate islet engraftment and prevent metabolic exhaustion and functional β-cell apoptosis; new immunosuppressive strategies better control islet graft rejection.

As a result we have seen a broader adoption of the islet transplantation technique and we have seen this therapy expand to be offered to many more patients with T1D [1]. This chapter focuses on the process of human islet cell isolation and its role in how to optimally provide cells for transplantation. However, with a great number of processes to outline, only the major ones will be focused on in this chapter. The major improvements in regards to the donor selection, islet isolation, transplantation and the immunosuppressive therapies used to improve outcomes to engraftment, function and survival of the islets will also be discussed. It is also acknowledged that there still remains the need for further ongoing

Table 6.1 A list of potential exclusion criteria that are suggested to be screened for in the donor [7]

[a]Bacterial	[b]Tuberculosis (TB), [b]Leprosy, [b]Treponema pallidum (Syphilis) and [b]multi resistant bacteria
[a]Fungal	[b]Cryptococcus, [b]Aspergillus
Viral	Human immunodeficiency virus (HIV), Hepatitis B (HBV), Hepatitis C (HCV), cytomegalovirus (CMV), Epstein Barr Virus (EBV), Zika virus
Parasitic	Malaria, Chagas disease, Schistosomiasis, and *Strongyloides*
Prion –	Creutzfeldt-Jacob disease
General exclusions	Agranulocytosis, aplastic anemia
Previous or current malignant neoplasms	Specifically melanoma and haematologic malignancies

[a]Positive bacterial and fungal cultures are a common occurrence following the donor procedure. These are usually due to the surgical process including bowel stapling and dissection to remove duodenum en-bloc with the pancreas. A high percentage of these positive cultures are from the skin and gut flora and are not an unexpected culture result. Part of processing the pancreas in the isolation laboratory is the decontamination of the pancreas, which is discussed later in this chapter

[b]The specific exclusions to this are the above mentioned positives for these bacterial and fungal pathogens

improvements to islet cell isolation and transplantation, but as a whole, islet cell transplantation offers a safe and feasible therapeutic option for an increasing number of patients suffering from T1D with severe hypoglycaemic unawareness [4].

The focus of clinical islet isolation lays in its ability to reliably provide islet cells that are of a sufficient number, viability and overall quality to allow the islet preparation to reach release criteria for transplantation every time the islet isolation process is undertaken. Underpinning this entire process is the reliance on the quality of the donor organ, which needs to be of a suitable size and quality to allow the production of enough viable islet cells. In order to undertake the isolation process for clinical transplantation, it is imperative that the donor pancreas be free from viral, bacterial, fungal, prion, cancer or genetic disease from the donor. Table 6.1 provides a guide to some of the more commonly occurring diseases that should be avoided when screening the donor before accepting the pancreas for islet cell isolation and subsequent clinical transplantation. Infectious risk factors depend on the history of the underlying disease of the transplanted organ, the donor, and the immunosuppressive treatment [5]. All pathogens, bacteria, viruses, fungi and parasites are possible but their frequency varies according to the transplanted organ, the selected immunosuppressive therapy and prophylaxis [6]. Obviously, there are many more variables with regards to donor selection criteria, and these will be discussed in more comprehensive detail in the following section on donor selection.

Careful selection and adherence to sterile procedures also flows through to processing of the pancreas tissue, and this includes the importance of taking microbiology culture samples at points throughout the entire isolation process to ensure protection to the recipient and regulatory compliance is met. These culture sample points are performed as interventional as well as precautionary as there remains the potential to unintentionally introduce contamination throughout the islet preparation process literally from the start to the completion of the process. This commences even prior to the processing of the organ with the organ donor and the surgery performed during the donor operation where there is the potential for exposure to a multitude of skin, gut and environmental pathogens despite adherence to the most stringent of sterile techniques [8, 9]. Less likely but still a potential point of unintentional contamination, is during the many steps undertaken throughout the isolation and the culture process prior to the transplant procedure [10, 11]. Along with this careful donor selection, there remain a significant number of other roadblocks and deviations that need be addressed to allow for islet cell isolation to be completed with a successful outcome, in order to be able to reach release criteria and allow safe and effective transplantation of the patient [12]. All islet preparations must be subjected to quality control assessment to reach a minimum standard to justify release and thus proceed to transplantation [13]. This is necessary to not only ensure the best possible outcome for recipients following transplant and to minimize chances of exposure to any tissue antigens or potentially transmissible agents from the donor, but also to justify exposing the patient to the unavoidable risks associated with the transplantation procedure such as anaesthesia and the need for life-long immunosuppression.

6.2 The Islet Processing Facility

The islet isolation facility is an integral part of this process, providing a regulated environment where the islets are purified from the donor pancreas. Clean room facilities of an appropriately legislated standard are compulsory to be able to provide a clean, safe and effective working environment for islet cell isolation or the manufacture of any biological or cellular product governed by the Regulatory Guidelines for Biologicals.

The clean room design minimizes the risk of contamination by providing an environment capable of limiting microbial contamination and lowers concentration of airborne particles to specified limits. These requirements are outlined in the ISO14644 documents for cleanrooms and associated controlled environments and can be found inthe International Standard ISO 14644 Cleanrooms and associated controlled environments [14]. Routine processing of cells or tissues in these types of cleanroom environment are run under the control of environmental programmes that provide monitoring and measurement of the cleanroom's performance. Eliminating submicron airborne contamination is a strictly regulated process with the level to which these particles need to be removed depending upon the government standards prescribed in each country.

However, as a general rule, control of contaminants generated by people, processes, facilities and equipment requires filtered air handling systems (utilising sealed systems with HEPA filters) that remove particulates from the air that enters the clean room. Effective contamination control is supported by appropriate clean room design, controlled air flow/direction, graded pressure, temperature, and humidity, effective validated cleaning systems and monitoring programmes. The Human Applications Laboratory cleanrooms at Westmead were designed and built utilising engineering expertise meeting prescribed requirements nominated in the Australian Code of Good Manufacturing Practice for human blood and blood components, human tissues and human cellular therapy products. As a guideline the recommended performance levels for the manufacture for products such as islets for clinical use are outlined in Table 6.2.

Clean rooms manufacturing biological products using sterile production also require strict aseptic gowning protocols designed to reduce the risk of any introduced contaminants or potential pathogens to the product. The use of face masks, full containment overalls, hoods, boots and gloves are used to reduce any potential for introduction of contaminants into the facility as can be seen in Fig. 6.1. All work undertaken within the facility is also performed within a Class II Biological Safety Cabinet. Along with the prevention of any potential contamination, the entire environment is strictly monitored and a battery of

Table 6.2 Shows the recommended limits for airborne contamination by particulates in the air for clean rooms both when not in use (at rest) and in use whilst carrying out product preparation. The maximum permitted of any type of particulate is given per m^3 equal to or greater than the tabulated size (Adapted from [14])

	Recommended limits for airborne contamination				
	Maximum permitted of particles per m^3 equal to or greater than the tabulated size				
	ISO	At rest		In operation	
Grade	EN14644-1	0.05 μm	0.5 μm	0.05 μm	0.5 μm
A	5	3,520	20	3,520	20
B	5	3,520	29	352,000	2,900
C	7	352,000	2,900	3,520,000	29,000
D	8	3,520,000	29,000	Not defined	Not defined

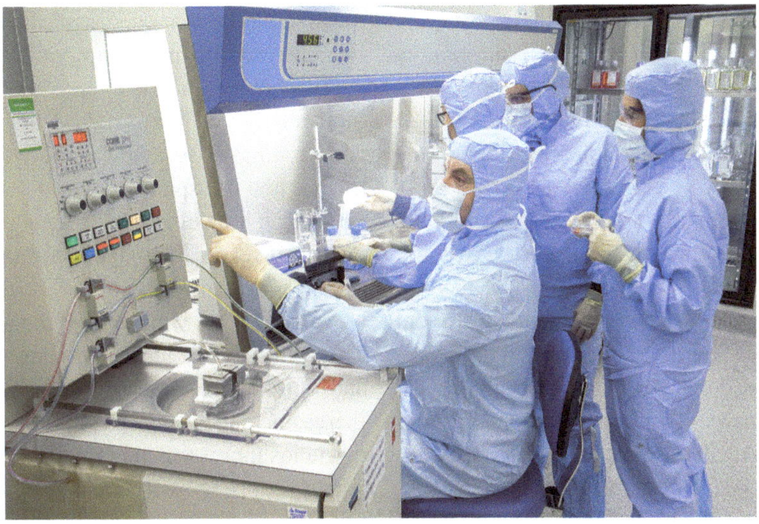

Fig. 6.1 A typical example of a clean room environment where pancreatic islet cells are isolated for clinical transplantation and can be also for research. Note the use of facemasks, full containment overalls, hoods, boots and gloves to reduce introduction of contaminants into the facility. All work is undertaken within closed systems and performed inside of Class II Biological Safety Cabinets as shown. The facility is the National Islet Transplant Laboratory within the Human Applications Laboratory (HAL) at the Westmead Institute for Medical Research on the Westmead Hospital campus, Westmead (Sydney), NSW, Australia

tests are performed to ensure close monitoring of the processes occur. Specifically, the environment is monitored where air samples are continuously collected throughout the duration of the isolation process by using air samplers, microbiological settle plates and contact plates of the work surfaces as outlined by the International Standard ISO 14644 Cleanrooms and associated controlled environments [14].

6.3 Clinical Islet Isolation Outcomes

We have seen significant improvements in outcomes achieved over the last decade and these have helped drive the increasing use of islet cell transplantation as a broadly applicable clinical therapy for treating T1D [15–17]. These improvements to outcome have resulted in islet cell transplantation being able to demonstrate successful outcomes with 92 % at 1-year and 83 % at 3-year survival rates. These are certainly as good as or better than the currently accepted gold standard of whole pancreas transplantation that has reported survival rates to be 80 % at 1 Year and 61 % at 3 Years post-transplant [1]. A number of in-depth analyses have shown that with advances in islet isolation and post-transplant management, islet transplantation does produce outcomes on par with whole pancreas transplantation and represents a clinically viable option to achieve long-term insulin independence in selected patients with T1D [18]. Governmental and health insurance providers in several countries now reimburse islet transplantation as part of clinical care. As the safety of the procedure and of adjunctive immunosuppressive therapies improve, and benefit accrues over potential risk, islet transplantation will be offered earlier in the course of the disease to these patients [19].

This chapter provides an overview of the processes involved with this extremely complex and intricate technology, which involves cutting edge technical means to ensure reliable, and consistent islet isolation outcomes that provide safe clinical transplantation for patients with minimal risk from the donor, associated islet cell isolation or the transplant procedure itself.

Despite the issues involved being complex and somewhat difficult to achieve in all centres undertaking islet cell isolation, they provide a clear platform to advance the treatments for T1D and to provide targeted therapies for the various groups of patients who suffer from T1D. At present, this therapy remains targeted at those patients who suffer from uncontrolled hypoglycaemic unawareness. This is due to a number of factors; firstly due to the low numbers of organ donors available and as such relatively few donor pancreata available for use, let alone the overall quality of the organs available for donation meaning that not all pancreata are suitable for islet isolation due to donor factors that will be further discussed in this chapter [20]. Secondly, the population currently targeted with treatment arguably show greatest benefit due to its life saving reversal of their hypoglycaemic unawareness following transplantation [21]. The significant benefits to both the patient and the whole of the community have been shown with comprehensive analysis demonstrating great benefits to both the recipient with prevention of hypoglycaemia or at worst death on the waiting list. In addition, reduction in hospitalization rates and associated medical costs following transplantation significantly lowers the overall cost to society, with studies showing that islet cell transplantation is clearly more effective than standard insulin treatment over the longer term [22].

Beckwith et al. performed a cost-effectiveness analysis and made a comparison with standard insulin therapy, using Markov modelling and Monte Carlo simulations. They showed that insulin therapy, cumulative cost per patient during a 20-year follow-up was $663,000, and cumulative effectiveness was 9.3 quality-adjusted life years (QALY), the average cost-effectiveness ratio being $71,000 per QALY. Islet transplantation had a cumulative cost of $519,000, a cumulative effectiveness of 10.9 QALY, and an average cost-effectiveness ratio of $47,800. During the first 10 years, costs for transplantation were higher, but cumulative effectiveness was higher from the start onwards. In sensitivity analyses, the need for one instead of two transplants during the first year did not affect the conclusions, and islet transplantation remained cost saving up to an initial cost of the procedure of $240,000. Their study showed that islet cell transplantation is more effective than standard insulin treatment, and becomes cost saving at about 9–10 years following first transplant.

We can now offer our patients with type 1 diabetes and severe hypoglycaemic unawareness the option of treatment with a clinical allo-islet cell transplant as a means to cure their diabetes [16]. This cutting edge technology has been available and utilized for over two decades with ever increasing success [23]. As the technologies and immunosuppressive therapies advance, the outcomes improve with greater options available and the functional survival rates also greatly improving to now be equivalent to those offered by whole organ pancreas transplantation rates [19].

A significant advantage of islet allograft transplantation is that it is a minimally invasive procedure to the recipient with the transplant being able to be performed by a number of relatively simple transplant methods. These are based around two main types of procedures usually being undertaken as a percutaneous transhepatic radiological procedure [24] or as a minilaparotomy and cannulation of a mesenteric vein to access the porta in order to infuse the islets into the liver [16].

Despite some units having selected success with single donor transplants [23, 25] it does however usually require two or more transplants to achieve insulin independence [15] with some units transplanting as many as five islet preparations into a single recipient, but this is not the norm. The greatest success remains with the fact that with just one transplant most patients become C-peptide positive and this has been extremely successful in reducing the underlying issue in these patients which is to prevent their severe hypoglycaemic episodes [26], making this a life-saving form of treatment for these patients which few current other therapies can offer.

Like its forerunner, simultaneous pancreas kidney transplantation, islet cell transplantation has been demonstrated to provide excellent success rates with improving long-term outcomes being shown to prevent ongoing progression of the other secondary complications of diabetes [27] and a number of studies have shown significant improvements in the secondary complications including retinopathy [13, 28, 29] and neuropathy [29].

This chapter provides a comprehensive outline to the methods currently available to improve the outcomes for islet isolation to ensure a treatment and as such cure for T1D for our patients. This also provides an ongoing advancing platform to base and develop the newer methods so that we can move forward with cutting edge technologies such as xeno-islet cell transplantation to provide a wider reaching treatment strategy for all patients suffering from type 1 diabetes.

6.4 The Donor Organ

The most significant hurdle to increasing the number of islet transplant recipients still remains the ongoing reliance on the extremely altruistic but still nevertheless low donation rate from cadaveric organ donors [30]. Improved donor management, organ recovery techniques, implementation of more stringent donor criteria, and improved islet cell processing techniques can contribute to enhance organ utilization for transplantation [31].

The problems associated with low organ donation rates are universal and are ongoing despite the best attempts to improve these with campaigns to educate and inform people of the benefits. However, we have seen recent improvements in organ donor rates and the uptake of methods of organ donation such as the use of deceased cardiac death (DCD) and utilisation of more marginal organ donation to expand the available donor numbers [9]. The current treatment rate for patients with T1D by these methods has seen improvement but despite this increase they ultimately remain comparatively low [32]. To be able to increase treatment rates for a greater proportion of patients, we rely on a focus on better utilisation of the current donor organs to provide improvements [9]. However, even with major increases in organ donor rates by such methods, we remain unlikely to be able to transplant the ever-increasing number of patients that require treatment for their T1D and development of their secondary complications such as renal failure [33]. Despite these issues, the current technology platform leads the way for treatment of patients with T1D by islet cell transplantation providing ever increasingly improved outcomes for patients in these sub-populations [1]. It also provides a means upon which we can base future advances in cellular therapies such as xenotransplantation to treat a broader range of patients that suffer from T1D in the future [34].

The donor therefore remains the integral factor influencing the success of isolation as this has significant impacting variables that affect the outcome of the islets for release at the end of the isolation procedure [35]. The donor organ contains the islets that are affected by the donor and their cofactors that the donor has been subjected to both genetically and environmentally. In regards to both genetic and environmental factors we must clearly screen the donor to ensure that the donor pancreas is free from viral, bacterial, fungal, prion, cancer or genetic disease. Table 6.1 provides a guide to some of the more commonly occurring diseases that should be avoided when screening the donor before accepting the pancreas for islet cell isolation and subsequent clinical transplantation. One of the original landmark studies in this area by Lakey et al. showed that there were critical factors in the multiorgan cadaveric donor that play an overall role in islet isolation outcomes which they identified using univariate analysis [36]. They identified a number of contributing co-factors that included donor age, body mass index (BMI), cause of death, and prolonged hypotensive episodes (systolic blood pressure <90 mmHg or mean arterial pressure <60 mmHg for > 15 min) requiring high vasopressors (>15 microgram/kg/min dopamine or >5 microgram/kg/min Levophed). In their independent analysis of 19 donor variables using multivariate logistic stepwise regression, they showed the most significant factors that were statistically significant and had an odds ratio (OR), demonstrating that donor age (OR 1.1, $P<0.01$), high BMI (OR 1.4, $P<0.01$), and local procurement team (OR 10.9, $P<0.01$), had a highly positive correlation with islet recovery. These studies eventually lead to the development of a score system to help aid in the subjective assessment of potential pancreata for islet isolation. Scoring systems are useful as a guide and as such follow essentially the variables described in this section of the chapter such as; Donor Age, BMI, CIT, and the procuring surgical team and techniques used to retrieve the pancreas.

We and others have subsequently also shown there are crucial factors that affect the isolation outcome due to the donor pancreas which include; the donor age, size (height and weight), BMI and overall health which play significant roles as does the cause of their hospitalisation and subsequent reason for their resulting donation. The direct admission and intensive care treatment including the length of time that they are managed prior to organ donation also plays a role in the isolation outcome. We have seen significant improvements to donor management and implementation of more stringent donor criteria that have contributed to enhance organ utilisation and thus transplantation [9, 31]. But despite these improvements, without care and improvement to organ retrieval by surgical teams trained and dedicated specifically to undertake pancreas retrieval as part of the multiorgan donor retrieval, we can see these improvements negated by the organ donor retrieval process [37]. Thus the appropriate choice of donor is imperative to provide a pancreas that can provide an islet cell preparation that can be transplanted safely and effectively into a recipient. These many factors have been investigated in depth over many decades and the most significant factors investigated are outlined here [23, 35, 38].

6.4.1 Effect of Donor Age

The first variable factor having been shown to play a significant role in isolation outcome by a number of studies is that of the donor age. It has been shown to significantly influence the outcome of the islet isolation process, this obviously contributes on its own but also in combination with other factors that will be discussed including their various interactions and causative outcomes. The age of the donor [39] has a significant impact for many reasons both due to the size of the organ but also from the perspective of how to undertake the digestion process. The influence it has on the overall outcome relies on the way the tissue is digested. Ultimately the amount of fat, vascular and connective tissue and percentage of fibrosis in the organ determine the speed and amount the surrounding tissues are digested and thus release the islet cells from them [40]. Obviously the easiest demonstration of the effect of age is by analysing the various age groups, and this is best observed by looking at the two major outliers; donor pancreata that are younger (<20 years of age) or older (>65 years of age), as they pose their own individual issues [40, 41]. Islet grafts isolated from young donors allow superior functional outcomes but are often associated with poor islet isolation yields with low numbers. The younger pancreata are obviously much smaller due to the donor's smaller size and overall weight and as such contain lower numbers of smaller islet cells. But this is not the only issue with younger donor pancreata; Meier et al. showed quite clearly that the pre-purification percentage of trapped or mantled islets was significantly higher in younger donors (44.3 ± 22.7 %) compared to >20 years of age donor pancreata (24.9 ± 20.9 %, $P<0.001$). This obviously leads to a lower recovery rate in younger donors (48 %) compared to >20 years of age donors (76 %, $P=0.002$) and hence results in lower post-purification islet equivalent (IEQ) per gram of pancreas in the younger donor ($2{,}412 \pm 1{,}789$ IEQ/g) compared to >20 years of age donor ($3{,}194 \pm 1{,}892$ IEQ/g, $P=0.01$). As a result the final islet cell yield is much lower in the younger donors at a mean of $180{,}982 \pm 128{,}073$ IEQ when compared to >20 years of age donors at $244{,}167 \pm 134{,}137$ IEQ, ($P=0.006$) [41]. A number of other studies have shown similar findings and have shown a strong negative correlation to isolation islet equivalent per gram (IEQ/g) pancreas in regards to younger donors (less than 20 years of age). As discussed this is due to the younger pancreas having more mantled and trapped islet cells and upon density separation undergoing significant losses of the islets due to their density being similar to the acinar tissue entrapping them. The islets and acinar tissue of the same density is pulled to the bottom of the density gradient used in separation of the islets from the acinar tissue, and so the islets are lost into the acinar and connective milieu. The same difficulties in separation of islets from acinar tissue in relation to age of the donor animal are also seen in animal models [42] and are described in greater detail in other chapters of this book.

At the other end of the age range are those donors deemed to be older with a number of studies having categorised islet donors into age groups with the general consensus being that organ donors 45 years of age or older provide overall better isolation results in terms of actual islet number and also IEQ/g of pancreas. However, there is a significant cost to using donors older than 45 years of age as they have negative outcomes in regards to transplantation due to a decrease of in-vivo function of the islets [43]. Niclauss et al. retrospectively analyzed 332 islet isolations according to donor age. In this study they investigated isolation outcome by islet yield, transplantation rate, and β-cell function in vitro. Transplanted patients were divided into two groups depending on donor age younger than or equal to 45 and older than 45-years of age. They showed that there was no difference in islet yields between the two groups (251,900 ± 14,100 and 244,600 ± 8,400 islet equivalent for ≤45- and >45-year-old donors, respectively). Transplantation rates and stimulation indices were similar in both groups as well. However, the significant differences were seen in the islet graft function parameters, which were significantly higher at 1-month follow-up in patients who had received islets from younger donors. At 6-month follow-up after second or third injection and at 12-month follow-up, secretory units of islets in transplantation indices and C-peptide/glucose ratios were significantly higher in patients with donors aged 45 years or younger.

Other studies have shown similar findings but have used slightly different age parameters with donors older than 50 years of age showing a worse outcome in regards to isolation outcomes. Cardillo et al. showed that there was a strong correlation with poorer isolation outcomes from those pancreas donors older than 50 years of age, with respect to the quality of the islet cells [39]. This was supported by data from the Edmonton group [36] showing that the insulin secretory capabilities of islets isolated from their >50-year-old donor group was significantly reduced as compared with the younger age group ($P < 0.02$). More interestingly to note is the fact that one of the few islet transplantation series reporting a consistent achievement of insulin independence after islet infusion from a single-donor have used strict donor selection criteria, limiting donor age to less than 50 years of age [44]. Obviously limiting the donor age creates significant issues with regards to reducing the potential organ donor pool and in our own unit, although our own preference is to utilise pancreata from organ donors younger than 50 years of age, we have used donor pancreases for islet isolation and subsequent transplantation from organ donors up to 63 years of age with good results [15].

These data suggest that, despite similar outcomes of the isolation procedure, islet graft function is significantly influenced by donor age. These results may have important consequences in the definition of pancreas allocation criteria. The major problem being that organ donors within the 20–45 years age group also happen to be the ideal donor age group for use as whole pancreas transplantation, and such is the case in a large number of simultaneous pancreas and kidney transplant programs [45]. At Westmead Hospital in Sydney, we see this directly as we run a National Islet Cell Transplant program alongside the National Simultaneous Pancreas and Kidney (SPK) Transplant program where preferential selection of donor pancreata prioritises younger aged pancreata to the SPK program. However, this can be viewed as a synergistic program as pancreata that would ordinarily be not utilised for clinical transplant such as from donor pancreases from donors 45 years or older or pancreases that have vascular disease, heavy fat infiltration, or fibrotic changes can be used for islet isolation rather than in the whole pancreas program.

A study from the Northern Italian group of Cardillo et al. analyzed the allocation protocols of all pancreas donors (2011–2012; n=433) in Northern Italy [39]. Outcome measures included donor characteristics and pancreas loss reasons during the allocation process. 23 % of the 433 pancreases offered for allocation were transplanted. Younger age, shorter ICU stay, traumatic brain death, and higher eGFR were predictors of pancreas transplant, either as vascularized organ or as islets. Among pancreas allografts offered to vascularized organ programs, 35 % were transplanted, and younger donor age was the only predictor of transplant. The most common reasons for pancreas withdrawal from the allocation process were donor-related factors. Among pancreas offered to islet programs, 48 % were processed, but only 14.2 % were transplanted, the most common reason for pancreas loss was due to unsuccessful islet isolation. Younger donor age and higher BMI were predictors of islet allograft transplant. As a result, they have changed the pancreas organ donor allocation strategy with equal distribution of donor pancreata between programs for either vascularized organ or islet transplant [39]. This is a practical means by which a number of programs around the world, including our own, can potentially improve transplant rates from the islet programs.

6.4.2 Effect of BMI

One of the original landmark studies in this area by Lakey et al. showed that there were clear benefits in using pancreata from larger donors but more importantly that higher body mass index (BMI) had a positive correlation with islet recovery [36]. The current literature provides strong evidence that continues to support this, with donors of BMI >25 providing significantly larger numbers of islets and significantly larger IEQ, leading to correspondingly higher chances of such isolations resulting in transplantation [12, 23].

Ponte et al. analyzed the effects of the donor and islet processing factors on the success rate of human islet cell processing for transplantation performed at their islet cell-processing center [31]. Higher islet yields were obtained from adult male donors, BMI >25 kg/m^2, showing adequate glycaemic control during hospital stay. Their data suggest that evaluation of the donor organ criteria prior to acceptance for islet isolation and ensuring the best possible criteria pancreas are selected is highly desirable to improve the success rate of islet cell processing. More recently, data from the Collaborative Islet Transplant Registry (CITR) report detailed the Islet Product Characteristics and Factors Related to Successful Human Islet Transplantation. One of the major findings was that donor body weight and BMI was associated with outcome of the IEQ count [23]. From this data, it would appear that the greater the BMI, the greater the number of IEQ. This is supported by some earlier studies that also evaluated significantly higher BMI donors such as BMI greater than 30 have even better IEQ. One such study by Sakuma et al. analyzed data from 207 islet isolations performed in their unit over a 5-year period with respect to donor characteristics, pancreas condition, and processing variables. They analysed the 207 isolations in regards to an outcome measure of more than 3,000 IEQ/g pancreas weight as being considered an acceptable isolation outcome. They showed a strong correlation with a positive outcome from donors with a BMI >30 kg/m^2 (P=0.002) [46].

However, there can be issues with larger BMI donors with a ceiling to larger BMI having a positive effect. There is a point at which a donor is too obese and pancreata are too heavily impregnated with fat to be beneficial instead they become problematic. As can be seen in (Fig. 6.2a, b), a donor pancreas for islet isolation heavily covered with a large amount of donor fat and connective tissues makes it difficult to cleanly dissect the pancreas free from the fat. It also makes it much more difficult to effectively decontaminate the pancreas for islet isolation which is essential to ensure a sterile product at the end of the isolation process which is described in detail in the next chapter. Also rather importantly the excess permeation of fat into the tissues requires greater distension of the gland for digestion and as such results in islets that are more fragile and

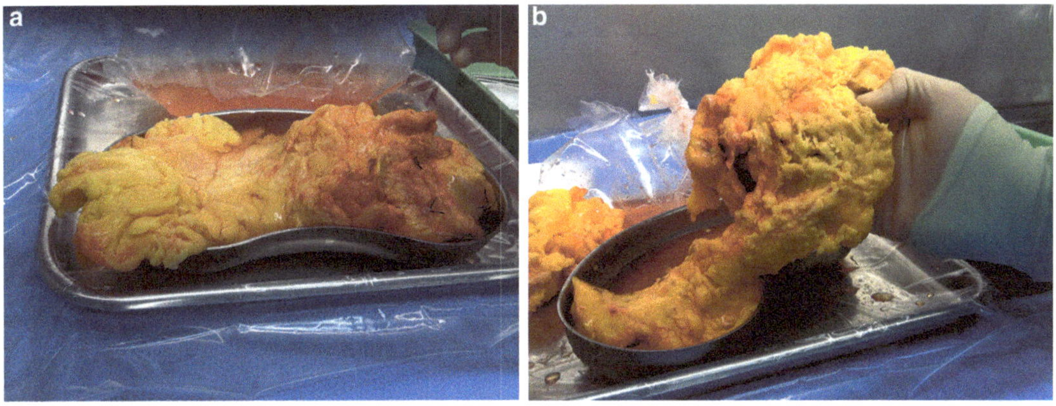

Fig. 6.2 (**a**) Shows a donor pancreas being unpacked from its transport bags as it is received into the isolation laboratory. This donor pancreas is heavily covered in donor fat and connective tissues, which make it difficult to cleanly dissect the pancreas free from the fat as in (**b**). It also makes it much more difficult to effectively decontaminate the pancreas for islet isolation

or readily destroyed in the isolation process. In fact, the amount of digestive enzyme used is even reduced to ensure that adequate dissociation occurs without over digestion and loss of islet cell numbers.

6.4.3 Effect of Cold Ischaemic Time (CIT)

Another very important but confounding factor that has an impact on isolation outcomes is the cold ischaemic time (CIT). The CIT is the time taken from the time of cessation of blood flow (cross clamp) in the organ donor and cold perfusion is commenced up until the time the pancreas is received into the processing laboratory and commences the islet isolation process. This is impacted by such variables such as surgical retrieval time, packaging, courier and transport times and receipting into the facilities. Being so multifactorial it is very dependent on logistical expertise of the organ donor network and systems they utilise. These are obviously quite variable dependent on the region in the world in which you live. In order to provide this service centralized islet isolation centers need to overcome a number of unique logistical problems, in particular retrieving donor pancreases and transplanting patients from distant areas. In Australia, this is particularly problematic [16] as our service covers an area of more than 7.5 million square kilometres, which is approximately twice the size of Europe or three-quarters the size of the United States. Almost one third of the Australian population lives outside these major urban centres and patients from regional and rural areas face a number of barriers to accessing medical services. So the importance of logistical expertise is paramount in order to minimise shipping times and ultimately CIT.

Traditionally the general consensus among clinical islet isolation and transplant centres is that a cold ischaemia time beyond 8 h results in significantly reduced yields and quality of human islets [47, 48]. A number of studies have previously shown that the longer the ischaemic time, the worse the isolation outcome in terms of IEQ and functional capacity of the islets. Wang et al. analyzed 276 islet isolations to identify variables for islet yield and, additionally, islet size and size distribution. Pearson correlation analyses demonstrated that CIT had a significantly negative correlation with actual islet count and islet equivalent (IEQ)/g (all $p \leq 0.003$) the longer the CIT the worse the outcome [12]. However, more recently studies have shown extension of the 8 h limit to extend this to as far as 12 h in order to be able to provide greater numbers of pancreata to process for islet isolation and potential transplantation and others have suggested extending this even further for research-only isolations [48, 49]. Kühtreiber et al. examined the isolation process for pancreata with extended CIT pancreata (mean of 13.2±0.7 h) and concluded that human islet isolation process permitted the recovery of large numbers of high-quality human islets from extended CIT pancreata [48]. More recently Lyon et al. examined the feasibility of a research-only human islet isolation and whether key criteria such as CIT and metabolic status may be relaxed and still allow successful research-focused isolations. They examined 142 isolations over approximately 5 years and confirmed that CIT had a negative impact on isolation purity and yield, and extending CIT beyond the typical clinical isolation cutoff of 12 h (to ≥ 18 h) had a modest impact on islet function [49].

However, the findings of the most recent Collaborative Islet Transplant Registry (CITR) data reported by Balamurugan et al. showed that only a limited number of the clinical islet isolations used pancreata extended past 10 h with the mean CIT of 9.3 h in isolations performed between 2007 and 2010 [23]. In our islet transplant program we prefer to not utilise pancreata with a CIT of greater than 10 h for isolation for clinical transplantation but will process for research preparations from pancreata with a CIT of more than 12 h.

To help aid in the subjective assessment of potential pancreata for islet isolation scoring systems are useful as a guide and as such follow essentially the variables described in this section of the chapter such as; Donor Age, BMI, CIT, procuring surgical team, cause of death, length of hospital stay, use of Vasopressors, social history, medical history, and other additional co-factors such as physical properties of the pancreas. These were originally described by O'Gorman et al. in a study that they undertook looking at all the potential variables that impacted their pancreas donor and islet isolation outcomes [35]. Developments of new score systems are showing resurgence as a means by which we can try to utilise a formula to provide a guide to accept donor pancreata for islet isolation. A score system is a useful exercise but ultimately it does not impact on the donor source as to what is available in your own country as this is entirely a reflection of the available donors and the ability of the donor agency to follow them to donation.

6.5 Organ Retrieval

As detailed previously there are a number of factors that can affect islet isolation outcomes, amongst these one of the most significant to effect islet isolation outcomes in regards to islet cell yield and function is the retrieval process and this is for a number of reasons. One of the strongest correlating factors was when the pancreas was retrieved by a surgical team from the isolation centre's hospital and this is most likely due to an obvious interest in retrieving the pancreas for islet isolation [36]. A number of subsequent papers have also demonstrated similar findings and a number of studies have shown direct effects from surgical retrieval of the pancreas during procurement. The effect using the local surgical team has on the pancreas retrieval outcome is to limit the amount of potential damage that can occur to the pancreas at retrieval. This impacts on the ability to adequately distend the pancreas upon injection of enzyme into the gland for distension and dissociation [37]. Quite clearly the surgical team involved in the donor organ retrieval plays a significant role and the UK Transplant Registry was analyzed to determine the frequency of pancreatic injuries, identify factors associated with damage, and assess the impact of injuries on graft survival. 1,296 pancreata were procured from donation after brain death donors. Surprisingly, more than 50 % of recovered pancreata had at least one injury. Following univariate analyses, they found the most important factors associated with increased rates of pancreas damage were from simultaneous liver donation, procurement team origin and increasing donor BMI. Damage to the pancreas during organ recovery is more common than other organs, and meticulous surgical technique and awareness of damage risk factors are essential to reduce rates of procurement-related injuries [50]. A number of "no touch" techniques

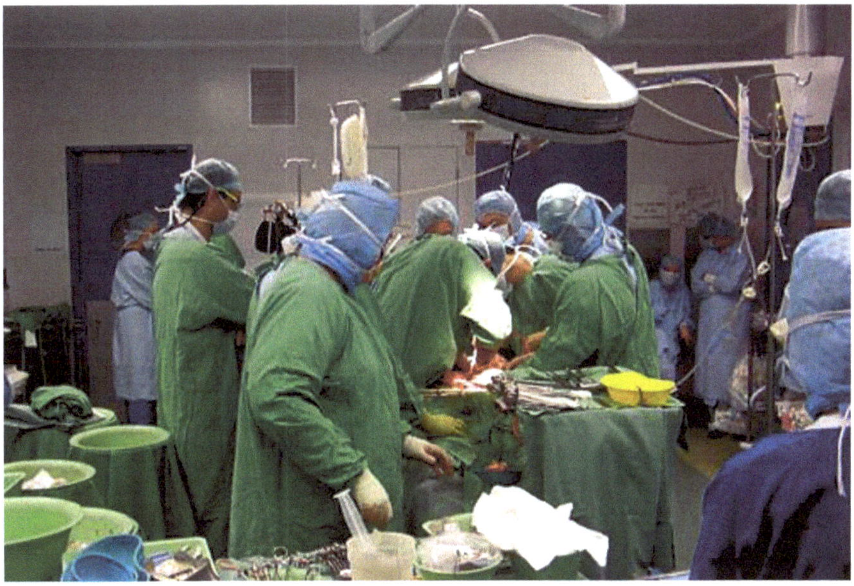

Fig. 6.3 Shows an example of a multiorgan abdominal surgical team working with the cardiothoracic surgical donor team performing an organ donor retrieval procedure at a major hospital, which is set up for multiorgan donor retrieval procedures. Note the great number of staff required to work effectively together as an integrated team to make the entire process occur as smoothly and flawlessly as possible

have been developed to ensure no damage occurs to the pancreas at retrieval including the use of the Harmonic scalpel to aid in the retrieval procedure as described by Hameed et al. [51]. Romanescu et al. also provide a very detailed description of the pancreatic retrieval procedure using a "no touch" technique where the spleen is used as a mechanical support or handle for pancreas mobilization for islet isolation [52]. Ensuring adherence to the no touch technique and ensuring that no damage to the capsule of the pancreas is essential to ensure complete distension when enzyme is infused to digest the pancreas. A study by Ponte et al. analyzed the effects of the donor and islet processing factors on the success rate of human islet cell processing for transplantation performed at their islet cell-processing center. Islet isolation outcomes improved with higher islet yields obtained when the local surgical team retrieved the pancreas. Their data suggest that a sequential, integrated approach including the use of a well-trained donor surgical team can improve the success rate of islet cell processing [31]. The Westmead Hospital transplant program has the advantage that it is a multiorgan donor retrieval service as well as a National Pancreas and a National Islet transplant unit. As such, it has vast experience in multiorgan retrieval for a period of decades with a dedicated focus on pancreas organ retrieval. This integrated approach to both multiorgan donation and transplantation has allowed a great focus from the time of donation to transplantation of the organ or isolation of islets. As can be seen in Fig. 6.3 the organ donor team must be versatile and able to readily integrate with all of the many other staff in the operating theaters to allow them to attend these procedures at any potential organ donor hospital. The donor team and the donor operation are significant factors in the overall outcome and quality of the islets. The importance of the organ retrieval procedure cannot be overstated as it has significant impacts on the outcomes of the islet isolation and as such is of the utmost importance to be done without damage to the pancreas and urgency the same as that of a pancreas that is being retrieved for whole pancreas transplantation.

6.6 Pancreas Preservation

As detailed in the previous section, the importance of the organ donor and the influence the donor pancreas plays on isolation outcome are also effected by the way the organ is retrieved in regards to the surgical team performing the retrieval and obviously the type of perfusate solution used for preservation of the donor pancreata. The currently most widely used perfusion technique is direct aortic flush of the organ donor with cold preservation solution which in the majority of units is still University of Wisconsin (UW) solution, however a number of other perfusion solutions have seen increasing use more recently with histidine-tryptophan-ketoglutarate (HTK) and Celsior (CS) being used along with other local agents in some centres. Most recently Balamurugan et al. reported for the CITR that of the total 1,017 pancreata retrieved for pancreas islet isolation between 2007 and 2010, over 42 % of abdominal donor organs perfused with UW solution, 7 % with HTK and 2.3 % with CS with the remainder not reported [23]. Since 2010 a number of other papers have shown predominance in the use of perfusion of the donor for multi-organ retrieval still with UW perfusion solution but that there is an increase in the use of HTK and CS to some degree [9, 53, 54].

Despite the apparently good outcomes with the use of UW solution a number of units have continued to investigate alternatives specifically for improving outcomes of the pancreas for islet isolation. However, this is very difficult, as there has to be a solution that covers all abdominal organs as once the perfusion commences, it perfuses all of the abdominal organs via the aorta at the same time. Some of the solutions that have been developed for this purpose have not essentially differed that greatly from the currently available UW, HTK, CS solutions and there remain concerns that supplementation of cold-storage solutions with cytoprotective agents and perfusion may improve pancreas and islet transplant outcomes [55].

A study evaluating the effects of one such solution was on Institut Georges Lopez-1 (IGL-1) a preservation solution similar to UW solution, however the ratio of Na/K are reversed. In a study by Niclauss et al. they assessed the impact of IGL-1, UW, and CS solutions on islet isolation and transplant outcome [56]. They retrospectively analyzed 376 islet isolations from pancreases flushed and transported with IGL-1 (n = 95), UW (n = 204), or CS (n = 77). Isolation outcome and β-cell function in vitro along with transplanted patients were divided into three groups depending on preservation solution of pancreas, and islet graft function was assessed by decrease in daily insulin needs, C-peptide/glucose ratios, β-scores, and transplant estimated function at 1- and 6-month follow-up. The IGL-1, UW, and CS groups were similar according to donor age, body mass index, and pancreas weight. There was no difference in islet yields between the three groups. Success rates, transplant rates, β-cell secretory function, and viability were similar for all three groups. They observed no difference in decreased insulin needs, C-peptide glucose ratios, β-scores, and transplant estimated function at 1- and 6-month follow-up between IGL-1, UW, and CS groups. Their study clearly showed that UW, CS and IGL-1 were equivalent solutions with no significant differences in outcomes for pancreas perfusion and cold storage before islet isolation and transplantation [56].

A number of units have reported differing methods for infusion or treatment of the pancreas and its storage/shipping of the organ at the time of retrieval, once the pancreas has been divided from the liver on the back table [36, 57]. Takita et al. have recently reported a study that evaluated the effects of two different solutions for pancreatic ductal perfusion (PDP) at organ procurement [57]. They studied the effects of treatment on 18 human pancreases assigned to three groups: non-PDP (control), PDP with ET-Kyoto solution, and PDP with cold storage/purification stock solution. Pancreatic islets were isolated according to the modified Ricordi method. No significant differences in donor characteristics, including coldischaemiatime, were observed between the three groups. All islet isolations in the PDP groups had more than 400,000 IEQ in total islet yield after purification, a significant increase when compared with the control

(P=0.04 and P<0.01). The islet quality assessments, including an in vivo diabetic nude mice assay and the response of high-mobility group box protein 1 to cytokine stimulation, also showed no significant differences. The proportion of terminal deoxynucleotidyl transferase (dUTP) nick-end labeling-positive cells showing apoptosis in islets in the PDP groups was significantly lower than in the control group (P<0.05). Both ET-Kyoto solution and cold storage/purification stock solution are suitable for PDP and consistently resulted in isolation success. These results appear to be encouraging, but further studies with a larger number of pancreas donors should be done to compare the effects of the PDP solutions [57].

6.6.1 The Two-Layer Method

Further studies have helped to develop other novel ways to treat the pancreas and improve its storage whilst shipping to the isolation centre. Kuroda et al. was first to report the so-called Two-Layer Method (TLM) in 1988 [58]. This method utilised a perfluorochemical (PFC) and initially Euro-Collins' solution, which was eventually replaced by UW solution to store the pancreas during shipping (Fig. 6.4a). The benefits of the use of the PFC are theoretically because it is a biologically inert liquid that acts as an oxygen-supplying media. A pancreas preserved using the TLM is oxygenated through the PFC and substrates are supplied by the UW solution. This allows the pancreas preserved using the TLM to generate adenosine triphosphate during storage, prolonging the preservation time [59]. The predominance of these methods has revolved around the use of oxygen exchange media such as the use of oxygenated fluorocarbons (Perfluorocarbon, Perfluorodecalin, perfluorohexyloctane and polydimethylsiloxane 5 (F6H8S5)) [59–61]. However, strong debate still remains of its benefits over the use of the static cold perfusion and storage using UW solution for preserving human pancreata prior to islet isolation [55, 61]. There are significant questions that relate to the oxygen exchange that can occur via the perfluorocarbon into the body of the pancreas whilst in the two-layer solution as can be seen in Fig. 6.4b where the pancreas sits upon the layer of perfluorocarbon with the UW solution on top. Despite this strong debate there has been a continuing level of research into the delivery of oxygen during solid organ preservation with the use of PFC. The one- and two-layer methods have been used as static storage techniques, proving popular for pancreas preservation trials. They have also been formulated as an emulsion for continual perfusion or as a simple flush solution. The success of PFC in

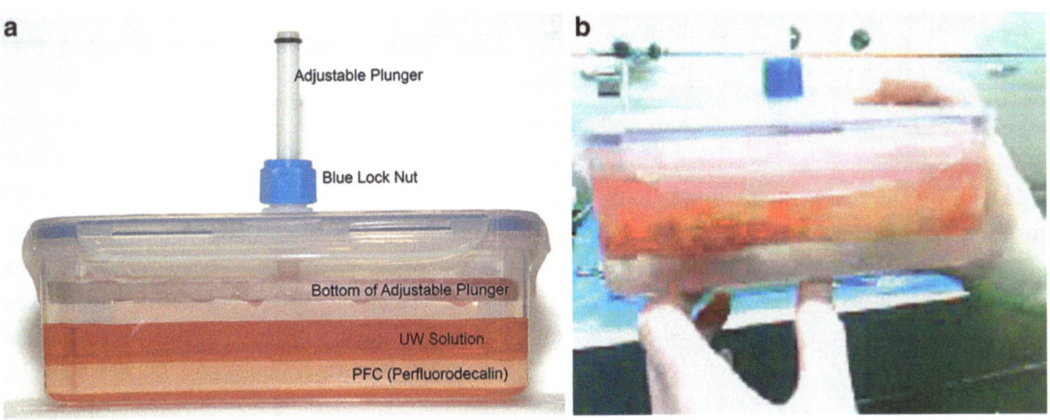

Fig. 6.4 (**a**) Shows an example of the type of pancreas transport device used for 2 layer storage for shipping from the donor hospital to the isolation facility. (**b**) Shows a perfused pancreas in the chamber showing the pancreas sitting on top of the UW/perfluorodecalin layers (Photo courtesy of Dr Tom Loudovaris, St Vincent's Institute, 9 Princes Street, Melbourne, VIC, 3065, Australia)

organ preservation seems to be somewhat organ and species dependent, and further experimental evidence is needed to establish their continued application [61, 62].

Recently, Li et al. performed a systematic review on donor characteristics and islet isolation outcomes between 2000 and 2013 [63]. They compared static UW perfusion alone with the TLM alone. From this they found that the TLM produced a significantly higher islet yield (weighted mean difference, 776.32; 95 % confidence interval; 370.82–1181.82; P=0.0002). TLM alone also yielded higher proportion of transplantable preparations (odds ratio, 1.60; 95 % confidence interval; 1.15–2.23; P=0.005). The following measures did not differ: islet viability (weighted mean difference, 2.10; −2.41-6.60; P=0.360), purity (weighted mean difference, −0.92; −3.75–1.91; P=0.520) and function assessed by measuring the stimulation index (weighted mean difference, 0.17; −0.21–0.55; P=0.380). When comparing TLM following UW storage with UW alone, the results were similar to the previous ones. These results indicate that the TLM can be used without detriment to islet yield and has the potential to increase the isolation outcomes resulting in improved human pancreatic islet transplantation rates [63].

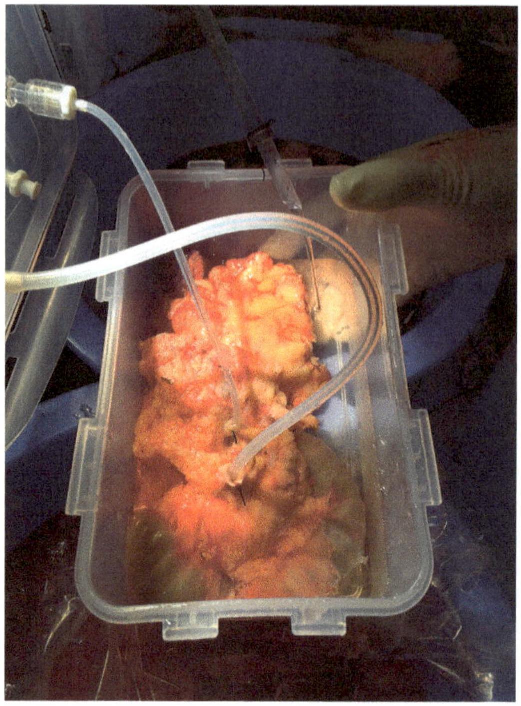

Fig. 6.5 Is a photo of machine perfusion being undertaken on a donor pancreas with cannulation of both the superior mesenteric and the splenic arteries to allow for complete perfusion of the pancreas with preservation fluid during transport to the isolation center (Photo courtesy of Dr Tom Loudovaris, St Vincent's Institute, 9 Princes Street, Melbourne, VIC, 3065, Australia)

6.6.2 Machine Perfusion

Like the advocates for the TLM, there are those that are now pursuing other options for pancreas preservation as they feel that the current methods have been identified as suboptimal due to insufficient oxygenation. Enhanced oxygen delivery is a key area of improvement. Scott and colleagues investigated other options for improving oxygen delivery to the pancreas whilst cold stored [64], such as by persufflation (PSF), i.e., vascular gas perfusion as can be seen in Fig. 6.5 where a donor pancreas is cannulated via both the superior mesenteric artery and the splenic artery to be able to perfuse the whole of the pancreas vasculature using the machine perfusion technique. In their study Scott and colleagues evaluated PSF on human pancreata obtained from brain-dead donors and also porcine pancreata procured by *en bloc* viscerectomy from heparinized organ donation in organ donors following cardiac death. In this study they performed a comparison of these pancreata where they were either preserved by the TLM or PSF. Following procurement, organs were transported to a 1.5-T magnetic resonance system for nuclear magnetic resonance spectroscopy to investigate their bioenergetic status by measuring the ratio of adenosine triphosphate to inorganic phosphate (ATP:P(i)) and for assessing PSF homogeneity by MRI. They clearly showed that both human and porcine pancreata can be effectively preserved by PSF. Under the MRI they also showed that pancreatic tissue was homogeneously filled with gas. TLM can effectively raise ATP:P(i) levels in rat pancreata but not in larger porcine pancreata. ATP:P(i) levels were almost undetectable in porcine organs preserved with TLM. When human or porcine organs were preserved by PSF, ATP:P(i) was elevated to levels similar to those observed in rat pancreata. The methods developed for human and porcine pancreas PSF homogeneously deliver oxygen throughout the organ. This elevates ATP levels during preservation and may improve islet isolation outcomes while enabling the use of marginal donors, thus expanding the usable donor pool [64].

This is obviously very important when we look at todays increased use of donation after cardiac death (DCD) donor organs for transplantation. Grafts from DCD are subjected to greater ischaemic insult and are at higher risk of poor functional outcome. Although conventional preservation techniques may be adequate for donation after brain death (DBD) and low-risk DCD pancreases, as the number of DCD pancreas transplants increase and the threshold for rejecting organs decreases, the importance of optimal preservation techniques is going to increase. The use of these newer techniques in pancreas preservation warrant further studies to ensure ongoing improvements to our potential pancreas donor organs for islet isolation [62].

6.7 The Pancreas

The process of organ donor selection, the organ donor operation and preservation obviously all play significant roles to underpin the overall success of the isolation process and these have all been clearly outlined in the previous sections. However, ultimately it is the pancreas and the various cofactors that have affected the organ in the overall processes undertaken to provide it as a donor pancreas for islet isolation. These also obviously affect its ability to yield islets that are suitable for safe and effective islet transplantation. To understand the various cofactors that may affect the organ it is also useful to understand the organ itself and herein is a description of the pancreas and how best to target our processes to provide higher yield of good numbers of viable islets in the isolation process.

The normal human pancreas is a soft vascularized organ contained within a fibrous capsule, the bulk of the pancreas is made up of 80–90 %exocrine (acinar) and 1–2 % endocrine (islet) tissue [65]. The islet cells can be seen very clearly as nests of endocrine cells on histology with the remainder of the organ being comprised of vascular, ductal, connective, fibrous and nerve tissues as seen in (Fig. 6.6a)

Although reliant on the size of the gland, there are approximately 1–1.5 million islets [66] in a healthy human pancreas, making up 1–2 % of the whole organ. Some studies report that the distribution of islets is similar between the head and body regions, but others have reported that they are up to two times higher numbers in the tail region [67]. However, a recent very detailed study by Ionescu-Tirgoviste et al. observed that 'islets are spread gradually from the head up to the tail section of the pancreas in the form of contracted or dilated islet routes' [68]. This requires confirmation by further studies as the analysis was performed on a single pancreas. We have seen fairly even distribution of the islets throughout the entire pancreas and it is not important in regards to the isolation process. The entire organ is used in the isolation process with the aim to remove as many islets from the surrounding acinar and connective tissues as possible.

Also understanding the ratio of the number of islets in proportion to the size of the pancreas is also an important issue. Understanding that there is a direct correlation of the size to the number of islets means that we know that the bigger the donor patient is and the bigger the donor pancreas is the more islets that we are likely to obtain are key in donor pancreas selection for isolation. The details of its size, form and make up should ideally be known to those about to perform the isolation process. Described as a solitary organ, Schaefer undertook autopsy and studies to describe the size and appearance of the human pancreata and found that in the cohort of pancreata studies, those from females on average weighed 84.88 ± 14.95 g and from males 90.31 ± 15.08 g (n=216) [69]. In a more recent study by Caglar et al., observations of male autop-

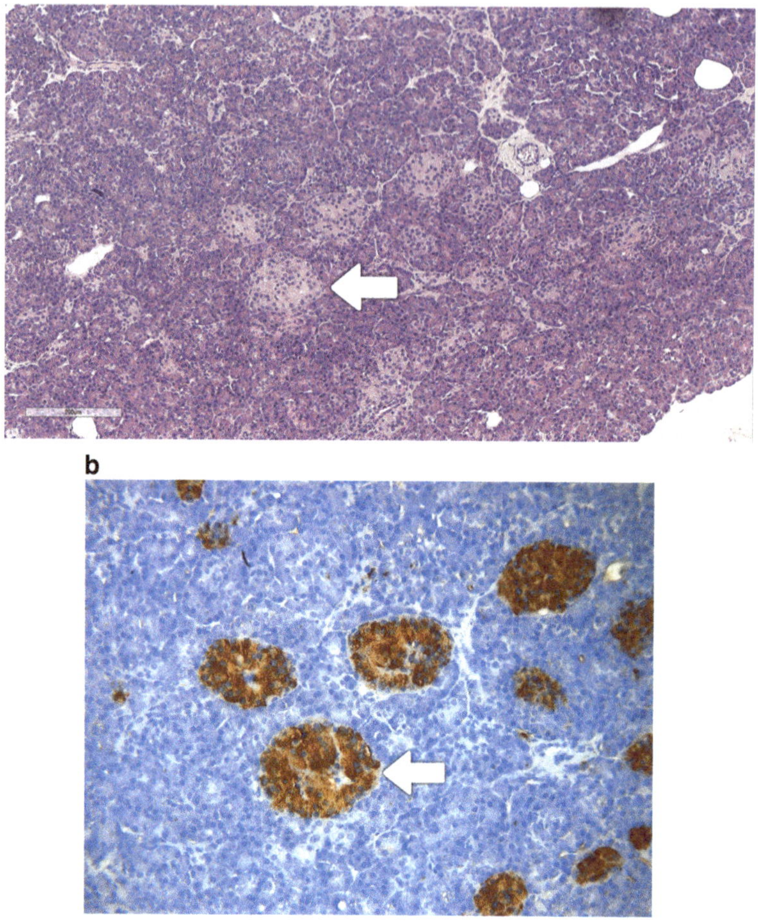

Fig. 6.6 (**a**) Shows an H&E stained histology section of a human pancreas at low power (100× mag). The islets can be seen stained *pink* (*white arrow*) surrounded by purple staining exocrine tissue. (**b**) A high power image (9200× magnification) of the same human pancreas section immunohistochemically stained for insulin, which appears as the brown areas as indicated by the *white arrow*

sies discovered similar findings with the average male pancreas to be 87.3 ± 30.6 g (range 41–174 g; n = 114), demonstrating significant variance in size between organs for potential islet isolation [70]. However of note is the fact that male donors tend to be on the whole larger than female donors and their corresponding pancreata correlate to this is important. Although due to the low organ donor numbers it is difficult to turn down any organ based on its size and certainly this is the case in our own program where regardless of the sex of the donor or their size we have been able to achieve successful isolation outcomes.

Islets are made up of endocrine cells involved in the secretion of hormones involved in the regulation of blood glucose levels. There are five types of endocrine cells: β cells, α cells, δ cells, γ or PP cells and ε cells (this is clearly seen in Fig. 6.6b) – where there is an example of immunohistochemistry staining of β cells in a human pancreas). According to Cabrera et al. and others, human islets are made up predominantly of ~60 % insulin secreting β cells and ~30 % glucagon secreting α cells [71, 72]. The somatostatin secreting δcells make up <10 %, pancreatic polypeptide secreting γ cells <5 % and a small number of ghrelin secreting ε cells. Laser scanning confocal microscopy used to generate serial optical sections through entire isolated human islets demonstrated that α and δ cells were intermingled with β cells in a heterogeneous manner [71]. This study also showed that β cells were common on the surface of isolated islets. This is in contrast to the core-mantle architecture of mouse and rat islets where α and δ cells are located around the edge of islets and β cells in the core. There are also non-endocrine cells within an islet, which include nerves, dendritic cells, macrophages, fibroblasts and vascular cells [73].

The vessels, which surround and penetrate the islets, deliver nutrients and oxygen to the islet cells and transport secreted hormones to target sites. This dense capillary network in which islets are embedded also provides a basement membrane, which is critical for islet function and viability [74, 75]. Ducts link the acinar tissue to the main pancreatic duct and possibly an accessory duct that deliver digestive enzymes into the duodenum. The ducts play a very important part in islet isolation as they enable delivery of the enzyme used to break down the pancreatic tissue and release islets. Connective tissue provides support for the pancreatic components and nerve tissue (sympathetic and parasympathetic) is involved in normal function.

In order to obtain large numbers of high quality human islets, the composition of the tissue, which surrounds islets, should be known. This aids in the correct choice of enzyme to enable release of islets from the surrounding tissue and ensure that islet integrity and vascularity is maintained. It also assists in the development of new enzymes to better suite this purpose. An immunohistochemical study by Hughes et al. identified and quantified the collagen subtypes of the islet-exocrine interface in the human pancreas from older (mean age 55.7 ± 3 year) and younger donors (21.8 ± 3.2 year) [76]. They reported that collagen VI was a major component of the adult human pancreas and was more than double that of collagen I or IV. Interestingly, they found that “the proportional collagen VI content was not dependent on age of the donor” so therefore may not be the reason mantling of islets occurs with younger donors. Regardless, such information can assist in the development of new enzymes with specific targets to improve clinical islet isolations, which have also started to be investigated in depth by others.

6.8 Concluding Remarks

In this chapter we have outlined the numerous advances in the techniques in islet transplantation outcomes by improvements to organ donor selection and management, organ perfusion, preservation and surgery, along with the facilities now necessary to provide the safe and effective isolation, culture and transplantation of clinical islets. In addition, the adoption and modification of newer immunosuppressive treatments have also helped to provide significant improvements in transplantation results and an overall increase to long-term outcomes [77]. We have also seen insulin independence rates of up to 10 years post-transplant with minimal complications [78] and the use of single pancreas donor infusions producing good long-term success [44]. However, islet transplantation still has limited application to the broader population of patients with T1D due to its reliance on the availability of cadaveric donor availability and selection, isolation results and transplant engraftment, the side effects of immunosuppression and issues associated with the requirement for life-long immunosuppression [79]. With further ongoing research in experimental and clinical studies we can see rapid improvements to islet transplantation which has moved from an experimental procedure to an effective clinical treatment option being offered now in a large number of islet transplant centres around the world offering another viable option to treat patients suffering from type 1 diabetes.

References

1. Barton FB, Rickels MR, Alejandro R, Hering BJ, Wease S, Naziruddin B, et al. Improvement in outcomes of clinical islet transplantation: 1999–2010. Diabetes Care. 2012;35(7):1436–45.
2. Radomski M, Zureikat A. Total pancreatectomy and islet cell autotransplantation: outcomes, controversies and new techniques. JOP. 2015;16(1):1–10.
3. Farney AC, Sutherland DE, Opara EC. Evolution of islet transplantation for the last 30 years. Pancreas. 2016;45(1):8–20.
4. Bertuzzi F, Antonioli B, Tosca MC, Galuzzi M, Bonomo M, Marazzi M, Colussi G. Islet transplantation in pediatric patients: current indications and future perspectives. Endocr Dev. 2016;30:14–22.
5. Greenwald MA, Kuehnert MJ, Fishman JA. Infectious disease transmission during organ and tissue transplantation. Emerg Infect Dis. 2012;18(8):e1.
6. Martinez-Pourcher V. Infections in the transplant patient. Rev Prat. 2015;65(8):1075–8.
7. Kovacs CS, van Duin D, Guedes de Morais A, Gonzalez BE, Avery RA, Mawhorter SD, Brizendine

KD, Cober E, Miranda C, Shrestha RK, Teixeira L, Mossad SB. Selecting suitable solid organ transplant donors: reducing the risk of donor-transmitted infections. World J Transplant. 2014;4(2):43–56.
8. Johnson CN, Morgan KA, Owczarski SM, Wang H, Fried J, Adams DB. Autotransplantation of culture-positive islet product: is dirty always bad? HPB (Oxford). 2014;16(7):665–9.
9. Shahrestani S, WA, Lam VW, Yuen L, Ryan B, Pleass HC, Hawthorne WJ. Outcomes from pancreatic transplantation in donation after cardiac death: a systematic review and meta-analysis. Transplantation. 2016; Epub Date 2016/03/08, PMCID: 26950713. doi:10.1097/TP.0000000000001084
10. Kin T, Rosichuk S, Shapiro AM, Lakey JR. Detection of microbial contamination during human islet isolation. Cell Transplant. 2007;16(1):9–13.
11. Murray L, McGowan N, Fleming J, Bailey L. Use of the BacT/alert system for rapid detection of microbial contamination in a pilot study using pancreatic islet cell products. J Clin Microbiol. 2014;52(10):3769–71.
12. Wang Y, Danielson KK, Ropski A, Harvat T, Barbaro B, Paushter D, Qi M, et al. Systematic analysis of donor and isolation factor's impact on human islet yield and size distribution. Cell Transplant. 2013;22(12):2323–33.
13. Maffi P, Secchi A. Clinical results of islet transplantation. Pharmacol Res. 2015;98(Aug):86–91.
14. U.S. Department of Health and Human Services FDA. International standard ISO 14644 and associated controlled environments [Available from: http://drug.fda.moph.go.th/drug/zone_gmp/files/GMP2549_2/Aug2106/7.ISO14644.pdf.
15. O'Connell PJ, Hawthorne WJ, Holmes-Walker DJ, Nankivell BJ, Gunton JE, Patel AT, et al. Clinical islet transplantation in type 1 diabetes mellitus: results of Australia's first trial. Med J Aust. 2006;184(5):221–5.
16. O'Connell PJ, Holmes-Walker DJ, Goodman D, Hawthorne WJ, Loudovaris T, Gunton JE, et al. Multicenter Australian trial of islet transplantation: improving accessibility and outcomes. Am J Transplant : Off J Am Soc Transplant Am Soc Transplant Surg. 2013;13(7):1850–8.
17. Robertson RP. Islet transplantation for type 1 diabetes, 2015: what have we learned from alloislet and autoislet successes? Diabetes Care. 2015;38(6):1030–5.
18. Moassesfar S MU, Frassetto LA, Szot GL, Tavakol M, Stock PG, Posselt AM. A comparative analysis of the safety, efficacy, and cost of islet versus pancreas transplantation in nonuremic patients with type 1 diabetes. Am J Transplant: Off J Am Soc Transplant Am Soc Transplant Surg. 2015.
19. Shapiro AM. State of the art of clinical islet transplantation and novel protocols of immunosuppression. Curr Diab Rep. 2011;11(5):345–54.
20. ANZDATA. Australia and New Zealand dialysis and transplant registry 38th report, chapter 2: prevalence of end stage kidney disease. 2016 [Available from: http://www.anzdata.org.au.
21. Shapiro AM. Islet transplantation in type 1 diabetes: ongoing challenges, refined procedures, and long-term outcome. Rev Diabet Stud : RDS. 2012;9(4):385–406.
22. Beckwith J, Nyman JA, Flanagan B, Schrover R, Schuurman HJ. A health economic analysis of clinical islet transplantation. Clin Transplant. 2012;26(1):23–33.
23. Balamurugan AN, Naziruddin B, Lockridge A, Tiwari M, Loganathan G, Takita M, et al. Islet product characteristics and factors related to successful human islet transplantation from the Collaborative Islet Transplant Registry (CITR) 1999–2010. Am J Transplant : Off J Am Soc Transplant Am Soc Transplant Surg. 2014;14(11):2595–606.
24. Owen RJ, Ryan EA, O'Kelly K, Lakey JR, McCarthy MC, Paty BW, et al. Percutaneous transhepatic pancreatic islet cell transplantation in type 1 diabetes mellitus: radiologic aspects. Radiology. 2003;229(1):165–70.
25. Al-Adra DP, Gill RS, Imes S, O'Gorman D, Kin T, Axford SJ, Shi X, Senior PA, Shapiro AM. Single-donor islet transplantation and long-term insulin independence in select patients with type 1 diabetes mellitus. Transplantation. 2014;98(9):1007–12.
26. Tiwari JL, Schneider B, Barton F, Anderson SA. Islet cell transplantation in type 1 diabetes: an analysis of efficacy outcomes and considerations for trial designs. Am J Transplant : Off J Am Soc Transplant Am Soc Transplant Surg. 2012;12(7):1898–907.
27. Thompson DM, Meloche M, Ao Z, Paty B, Keown P, Shapiro RJ, Ho S, Worsley D, Fung M, Meneilly G, Begg I, Al Mehthel M, Kondi J, Harris C, Fensom B, Kozak SE, Tong SO, Trinh M, Warnock G. Reduced progression of diabetic microvascular complications with islet cell transplantation compared with intensive medical therapy. Transplantation. 2011;91(3):373–8.
28. Chow VC, Pai RP, Chapman JR, O'Connell PJ, Allen RD, Mitchell P, et al. Diabetic retinopathy after combined kidney-pancreas transplantation. Clin Transplant. 1999;13(4):356–62.
29. Usuelli V, La Rocca E. Novel therapeutic approaches for diabetic nephropathy and retinopathy. Pharmacol Res. 2015;98:39–44.
30. Hays R. Transplant community looks for ways to incentivize living organ donation. Nephrol News Issues. 2016;30(1:24):26–7.
31. Ponte GM, Pileggi A, Messinger S, Alejandro A, Ichii H, Baidal DA, Khan A, Ricordi C, Goss JA, Alejandro R. Toward maximizing the success rates of human islet isolation: influence of donor and isolation factors. Cell Transplant. 2007;16(6):595–607.
32. Cravedi P, van der Meer IM, Cattaneo S, Ruggenenti P, Remuzzi G. Successes and disappointments with clinical islet transplantation. Adv Exp Med Biol. 2010;654:749–69.
33. Morrissey PE, Monaco AP. Donation after circulatory death: current practices, ongoing challenges, and

potential improvements. Transplantation. 2014;97(3):258–64.
34. Basta G, Montanucci P, Calafiore R. Islet transplantation versus stem cells for the cell therapy of type 1 diabetes mellitus. Minerva endocrinologica. 2015.
35. O'Gorman D, Kin T, Murdoch T, Richer B, McGhee-Wilson D, Ryan EA, et al. The standardization of pancreatic donors for islet isolations. Transplantation. 2005;80(6):801–6.
36. Lakey JRWG, Rajotte R, Suarez-Aimazor ME, Ao Z, Shapiro AM, Kneteman NM. Variables in organ donors that affect the recovery of human islets of Langerhans. Transplantation. 1996;61(7):1047–53.
37. Andres A, Kin T, O'Gorman D, Bigam D, Kneteman N, Senior P, et al. Impact of adverse pancreatic injury at surgical procurement upon islet isolation outcome. Transplant Int : Off J Eur Soc Organ Transplant. 2014;27(11):1135–42.
38. Pepper AR, Gala-Lopez B, Ziff O, Shapiro AJ. Current status of clinical islet transplantation. World J Transplant. 2013;3(4):48–53.
39. Cardillo M, Nano R, de Fazio N, Melzi R, Drago F, Mercalli A, et al. The allocation of pancreas allografts on donor age and duration of intensive care unit stay: the experience of the North Italy Transplant program. Transplant Int : Off J Eur Soc Organ Transplant. 2014;27(4):353–61.
40. Balamurugan AN, Chang Y, Bertera S, Sands A, Shankar V, Trucco M, Bottino R. Suitability of human juvenile pancreatic islets for clinical use. Diabetologia. 2006;49(8):1845–54.
41. Meier RP, Sert IM P, Muller YD, Borot S, Badet L, Toso C, et al. Islet of Langerhans isolation from pediatric and juvenile donor pancreases. Transpl Int. 2014;27(9):949–55.
42. Nagaraju S, Bottino R, Wijkstrom M, Trucco M, Cooper DK. Islet xenotransplantation: what is the optimal age of the islet-source pig? Xenotransplantation. 2015;22(1):7–19.
43. Niclauss N, Bosco D, Morel PD-M S, Brault C, Milliat-Guittard L, Colin C, Parnaud G, et al. Influence of donor age on islet isolation and transplantation outcome. Transplantation. 2011;91(3):360–6.
44. Hering BJ, Kandaswamy R, Ansite JD, Eckman PM, Nakano M, Sawada T, Matsumoto I, Ihm SH, Zhang HJ, Parkey J, Hunter DW, Sutherland DE. Single-donor, marginal-dose islet transplantation in patients with type 1 diabetes. Jama. 2005;293(7):830–5.
45. Thwaites SE, Gurung B, Yao J, Kable K, Robertson P, Ryan BJ, Lam VW, Pleass HC, Chapman JR, Hawthorne WJ, Allen RD. Excellent outcomes of simultaneous pancreas kidney transplantation in patients from rural and urban Australia: a national service experience. Transplantation. 2012;94(12):1230–5.
46. Sakuma Y, Ricordi C, Miki A, Yamamoto T, Pileggi A, Khan A, Alejandro R, et al. Factors that affect human islet isolation. Transplant Proc. 2008;40(2):343–5.
47. Kaddis JS, Danobeitia JS, Niland JC, Stiller T, Fernandez LA. Multicenter analysis of novel and established variables associated with successful human islet isolation outcomes. Am J Transplant. 2010;10:646–56.
48. Kuhtreiber WM, Ho LT, Kamireddy A, Yacoub JA, Scharp DW. Islet isolation from human pancreas with extended cold ischemia time. Transplant Proc. 2010;42(6):2027–31.
49. Lyon J, Manning Fox JE, Spigelman AF, Kim R, Smith N, O'Gorman D, Kin T, Shapiro AM, Rajotte RV, MacDonald PE. Research-focused isolation of human islets from donors with and without diabetes at the Alberta diabetes institute IsletCore. Endocrinology. 2016;157(2):560–9.
50. Ausania F, Drage M, Manas D, Callaghan CJ. A registry analysis of damage to the deceased donor pancreas during procurement. Am J Transplant : Off J Am Soc Transplant Am Soc Transplant Surg. 2015;15(11):2955–62.
51. Hameed A, Yu T, Yuen L, Lam V, Ryan B, Allen R, et al. Use of the harmonic scalpel in cold phase recovery of the pancreas for transplantation: the Westmead technique. Transpl Int. 2016.
52. Romanescu D, Gangone E, Boeti MP, Zamfir R, Dima SO, Popescu I. Technical aspects involved in the harvesting and preservation of the pancreas used for pancreatic islet allotransplantation. Chirurgia (Bucur). 2013;108(3):372–80.
53. Cao YS, S., Chew HC, Crawford M, Macdonald PS, Laurence J, Hawthorne WJ, et al. Donation after circulatory death for liver transplantation: a meta-analysis on the location of life support withdrawal affecting outcomes. Transplant. 2016.
54. Dikdan GS, Mora-Esteves C, Koneru B. Review of randomized clinical trials of donor management and organ preservation in deceased donors: opportunities and issues. Transplantation. 2012;94(5):425–41.
55. Iwanaga YS, Sutherland DE, Harmon JV, Papas KK. Pancreas preservation for pancreas and islet transplantation. Curr Opin Organ Transplant. 2008;13(2):135–41.
56. Niclauss N, Wojtusciszyn A, Morel P, Demuylder-Mischler S, Brault C, Parnaud G, et al. Comparative impact on islet isolation and transplant outcome of the preservation solutions Institut Georges Lopez-1, University of Wisconsin, and Celsior. Transplantation. 2012;93(7):703–8.
57. Takita M, Itoh T, Shimoda M, Kanak MA, Shahbazov R, Kunnathodi F, et al. Pancreatic ductal perfusion at organ procurement enhances islet yield in human islet isolation. Pancreas. 2014;43(8):1249–55.
58. Kuroda YKT, Suzuki Y, Fujiwara H, Yamamoto K, Saitoh Y. A new, simple method for cold storage of the pancreas using perfluorochemical. Transplantation. 1988;46:457–60.
59. Fujino Y. Two-layer cold storage method for pancreas and islet cell transplantation. World J Gastroenterol : WJG. 2010;16(26):3235–8.

60. Brandhorst H, Asif S, Andersson K, Theisinger B, Andersson HH, Felldin M, Foss A, Salmela K, Tibell A, Tufveson G, Korsgren O, Brandhorst D. A new oxygen carrier for improved long-term storage of human pancreata before islet isolation. Transplantation. 2010;89(2):155–60.
61. Hosgood SA, Nicholson ML. The role of perfluorocarbon in organ preservation. Transplantation. 2010;89(10):1169–75.
62. Barlow AD, Hosgood SA, Nicholson ML. Current state of pancreas preservation and implications for DCD pancreas transplantation. Transplantation. 2013;95(12):1419–24.
63. Li X, Zhang J, Sang L, Chu Z, Dong M. Influence of the two-layer preservation method on human pancreatic islet isolation: a meta-analysis. Int J Artif Organs. 2015;38(3):117–25.
64. Scott WE, Weegman BP, Ferrer-Fabrega J, Stein SA, Anazawa T, Kirchner VA, et al. Pancreas oxygen persufflation increases ATP levels as shown by nuclear magnetic resonance. Transplant Proc. 2010;42(6): 2011–5.
65. Kim A, Kilimnik G, Guo C, Sung J, Jo J, Periwal V, et al. Computer-assisted large-scale visualization and quantification of pancreatic islet mass, size distribution and architecture. J Visualized Exp: JoVE. 2011(49).
66. Korsgren O, Nilsson B, Berne C, Felldin M, Foss A, Kallen R, et al. Current status of clinical islet transplantation. Transplantation. 2005;79(10):1289–93.
67. Wang X, Misawa R, Zielinski MC, Cowen P, Jo J, Periwal V, et al. Regional differences in islet distribution in the human pancreas – preferential beta-cell loss in the head region in patients with type 2 diabetes. PLoS One. 2013;8(6):e67454.
68. Ionescu-Tirgoviste C, Gagniuc PA, Gubceac E, Mardare L, Popescu I, Dima S, et al. A 3D map of the islet routes throughout the healthy human pancreas. Sci Rep. 2015;5:14634.
69. Schaefer JH. The normal weight of the pancreas in the adult human being: a biometric study. Anat Rec. 1926;32(2):119–32.
70. Caglar V, Kumral B, Uygur R, Alkoc O, Ozen O, Demirel H. Study of volume, weight and size of normal pancreas, spleen and kidney in adults autopsies. Forensic Med Anat Res. 2014;2:63–9.
71. Brissova M, Fowler MJ, Nicholson WE, Chu A, Hirshberg B, Harlan DM, et al. Assessment of human pancreatic islet architecture and composition by laser scanning confocal microscopy. J Histochem Cytochem : Off J Histochem Soc. 2005;53(9): 1087–97.
72. Cabrera O, Berman DM, Kenyon NS, Ricordi C, Berggren PO, Caicedo A. The unique cytoarchitecture of human pancreatic islets has implications for islet cell function. Proc Natl Acad Sci U S A. 2006;103(7): 2334–9.
73. Eberhard D, Kragl M, Lammert E. 'Giving and taking': endothelial and beta-cells in the islets of Langerhans. Trends Endocrinol Metab: TEM. 2010;21(8):457–63.
74. Otonkoski T, Banerjee M, Korsgren O, Thornell LE, Virtanen I. Unique basement membrane structure of human pancreatic islets: implications for beta-cell growth and differentiation. Diabetes Obes Metab. 2008;10 Suppl 4:119–27.
75. Virtanen I, Banerjee M, Palgi J, Korsgren O, Lukinius A, Thornell LE, et al. Blood vessels of human islets of Langerhans are surrounded by a double basement membrane. Diabetologia. 2008;51(7):1181–91.
76. Hughes SJ, Clark A, McShane P, Contractor HH, Gray DW, Johnson PR. Characterisation of collagen VI within the islet-exocrine interface of the human pancreas: implications for clinical islet isolation? Transplantation. 2006;81(3):423–6.
77. Nanji SA, Shapiro AM. Advances in pancreatic islet transplantation in humans. Diabetes Obes Metab. 2006;8(1):15–25.
78. Bruni A, Gala-Lopez B, Pepper AR, Abualhassan NS, Shapiro AJ. Islet cell transplantation for the treatment of type 1 diabetes: recent advances and future challenges. Diabetes Metab Syndr Obes : Targets Ther. 2014;7:211–23.
79. Tuduri E, Bruin JE, Kieffer TJ. Restoring insulin production for type 1 diabetes. J Diabetes. 2012;4(4):319–31.

Clinical Islet Isolation

7

Wayne J. Hawthorne, Lindy Williams,
and Yi Vee Chew

Abstract

The overarching success of islet transplantation relies on the success in the laboratory to isolate the islets. This chapter focuses on the processes of human islet cell isolation and the ways to optimally provide islet cells for transplantation. The major improvements in regards to the choice of enzyme type, way the digested pancreas tissue is handled to best separate islets from the acinar and surrounding tissues, the various methods of purification of the islets, their subsequent culture and quality assurance to improve outcomes to culminate in safe and effective islet transplantation will be discussed. After decades of improvements, islet cell isolation and transplantation now clearly offer a safe, effective and feasible therapeutic treatment option for an increasing number of patients suffering from type 1 diabetes specifically for those with severe hypoglycaemic unawareness.

Keywords

Diabetes • Insulin • Islet • Islet cell • Islet cell allotransplantation • Islet cell autotransplantation • Islet cell isolation • Islet equivalent (IEQ) • Type 1 diabetes (T1D)

W.J. Hawthorne (✉)
National Pancreas and Islet Transplant Laboratories, The Westmead Institute for Medical Research, Westmead, NSW 2145, Australia

Department of Surgery, Westmead Clinical School, Westmead Hospital, University of Sydney, Westmead, NSW 2145, Australia
e-mail: wayne.hawthorne@sydney.edu.au; http://www.westmeadinstitute.org.au

L. Williams • Y.V. Chew
National Pancreas and Islet Transplant Laboratories, The Westmead Institute for Medical Research, Westmead, NSW 2145, Australia
http://www.westmeadinstitute.org.au; http://www.westmeadinstitute.org.au

Abbreviations

BMI	Body mass index
BSE	Bovine spongiform encephalopathy
CMRL	Connaught medical research laboratories
IEQ	Islet cell isolation
IEQ/g	Islet equivalent islet equivalents per gram
MTC	Mixed treatment comparison
T1D	Type 1 diabetes
UW	University of Wisconsin solution

M. Ramírez-Domínguez (ed.), *Pancreatic Islet Isolation*, Advances in Experimental Medicine and Biology 938, DOI 10.1007/978-3-319-39824-2_7

7.1 Pancreas Digestion: Separating Islets from the Surrounding Tissue

7.1.1 Receipt of the Pancreas into the Laboratory

Following the organ donor operation, organ perfusion and cold shipping, the pancreas is taken to the clinical isolation laboratory to undergo isolation. In this process there are a series of integral steps undertaken to ensure the best possible outcomes in terms of number of islets, islet equivalents per gram (IEQ/g) pancreas, quality of the isolated islets and of course their functional capacity. However, first and foremost is safety to the recipient patient who is both diabetic and immunosuppressed due to the immunosuppressive drugs used for induction and ongoing suppression of the immune system. This allows the islets to not be rejected by the recipient's immune system but at the same time potentially poses inherent risks to the recipient due to a depressed natural immunity for protection against potential pathogens. As such it is integral to not introduce any such potential pathogens to the patient by way of the islet transplant. To do this we have developed a number of steps to minimise any potential introduction of pathogens to a recipient. These include multi step surveillance of the process by collection of samples that are processed for identification of any potential microorganisms throughout the isolation process and culture periods. Prior to release of the islets for transplantation, a strict release criteria protocol is followed and conformance to this includes both assessment for microbiological and any potential endotoxin burden in the islet preparation. However, the very first steps in the whole process are to minimise any potential introduction of pathogens into the process by way of decontamination and prevention of such.

Figure 7.1a shows a donor pancreas after receipt into the islet isolation facility. The pancreata are generally retrieved using the "No touch technique" and still have surrounding connective tissue, fat, blood vessels and quite often a segment of duodenum and or spleen still attached [1]. As mentioned, the organ is transported in sterile bags with the pancreas in cold preservation solution. When it arrives the first thing to be done is to collect a sample of the transport media for microbiological assessment (refer to Sect. 7.5.3). The extraneous tissue is removed prior to cannulation of the pancreatic duct to reveal the head, middle (body) and tail [2, 3] as can be seen in Fig. 7.1b.

To minimise the introduction of contaminants into the isolation process, the trimmed and cannulated pancreas is decontaminated through a series of 4 °C solutions (Fig. 7.1c). Using sterile instruments, the pancreas is removed from the UW perfusion/transport solution and placed in a 10 % Povidone-Iodine bath for a period of no less than 3 min before then being transferred to a bath of 2 mg/ml Cephazolin/M199 solution for a period of no less than 3 min before finally being rinsed in cold media such as M199 to remove the decontaminating solutions. This process is done immediately prior to the pancreas being transferred to a dish for enzyme distension as described in Sect. 7.1.3.

7.1.2 Enzymes for Digestion of the Pancreas

The release of islets from surrounding exocrine and connective tissue remains a limiting factor in the success of islet isolation and the resulting transplantation of islets to treat individuals with type 1 diabetes (T1D). To add to the complexity associated with the architecture of the pancreas, donor variables and retrieval conditions, each described previously, is the variability of enzyme types, methods of introducing the enzyme into the pancreas (manual vs machine) and shaking of the Ricordi chamber (automated vs manual; [4]).

Early attempts at obtaining large numbers of high quality islets from donor human pancreases were hampered by the use of crude enzyme blends. These were very often mixtures of various collagenases, proteases and unknown enzymes, which degraded many types of collagen [5]. This meant there was variability in enzyme activity and concentration between

Fig. 7.1 (**a**) Shows a pancreas received into the islet isolation laboratory. It is still in the organ perfusion fluid – UW solution – from the donor procedure and transport. Note the presence of connective tissue, blood vessels, fat and a segment of duodenum. (**b**) Shows the same pancreas from (**a**) after removal of all extraneous connective tissue, blood vessels, fat and the segment of duodenum. It is still in the UW organ perfusion fluid from the donor procedure about to undergo decontamination. In (**c**), the pancreas has been removed from the UW solution and is being taken through decontamination. The kidney dish on the left contains sterile Povidone-Iodine, the second containing Cephazolin, the last M199 media wash to remove any of the povidone and antibiotic mix

batches making it difficult to achieve reproducibility in islet isolation procedures.

It was hoped the release of Liberase HI, a standardized highly purified collagenase and thermolysin blend, would enable the achievement of consistency in clinical islet isolations. Early reports when it was compared to the traditional collagenase preparation used for islet isolation (collagenase P) were hopeful, as islet yield was significantly higher and no differences were observed in this outcome when different batches were used [6]. However, further studies showed that this enzyme blend was variable between and also within lots [7]. In spite of this, many Islet Isolation and Transplant Units, including our own, were able to transplant islets, isolated with Liberase HI, into individuals with T1D and hypoglycaemic unawareness [8–10]. In many cases, this resulted in the cessation or reduction of endogenous insulin and abolition of hypoglycaemic unawareness, even when insulin was required.

However, the use of Liberase HI for clinical islet isolation stopped in 2007 when it was revealed that the *Clostridium histolyticum* from which it was manufactured was cultured in a broth containing bovine brain and may have posed a risk of transmission of Bovine Spongiform Encephalopathy (BSE). The manufacturer, Roche Diagnostics Corp determined that “the risk of BSE prions being present in the bovine material used in the production of Brain

Heart Infusion broth, carried through the production for the Liberase Purified Enzyme Blends and subsequent isolation and purification of cells used in clinical applications is remote, less than 1 in 1 million probability." [11].

During the Liberase HI period, another collagenase enzyme and neutral protease supplement was being tested – Collagenase NB1 and Neutral protease NB (SERVA Electrophoresis GMbH). This combination was reported to yield a similar number of islets with similar glucose stimulation indices as Liberase HI [12]. Sabek et al. also demonstrated that there was no difference in islet yield, purity as well as in vivo function as assessed by transplantation into diabetic NOD-SCID mice, when Liberase was compared with the SERVA enzymes [13]. Results obtained from a retrospective study comparing islet isolations performed with Liberase HI and the premium grade of Collagenase NB1 suggested that although the former was more efficient in pancreas dissociation, the percentage of islet isolations that reached criteria for clinical transplantation was the same with both enzyme types [14]. Observations from both studies indicated that Collagenase NB1 and Neutral protease NB caused less damage to the islets and surrounding tissue and may have been associated with the higher purity and reduced apoptotic rate seen.

An advantage of these SERVA enzymes is that they are available as a GMP product i.e. they are manufactured in compliance with the EU guidelines for good manufacturing practice (GMP) and fulfilled the requirements of TSE guidelines according to the European Pharmacopoeia. As such, they are more suited for clinical use from a safety aspect. The combination of the GMP Collagenase NB1 and Neutral protease NB (SERVA enzymes) has been used by our Unit and others to successfully treat individuals with T1D and hypoglycaemic unawareness [15–17].

This also stimulated the development of a mammalian tissue-free enzyme by Roche Diagnostics – called Liberase MTF (mammalian tissue free). This is similar to Liberase HI, however, it is manufactured under GMP, and as the name suggests, in the absence of mammalian tissue. Initial tests carried out by groups in the Netherlands, Sweden and France used a pre-mixed blend of collagenase and thermolysin – Liberase MTF-S – resulting in 9 of 12 isolations reaching criteria for clinical transplantation [18]. Criteria for potential clinical transplantation were ≥250,000 islet equivalents, static stimulation ≥1 and ≥70 % fluorescent viability. There was a large variability in the results and the authors suggested that this was likely due to inexperience using the new enzyme blend and possible inability to adjust the thermolysin concentration individually.

Another study compared the islet isolation outcomes of Liberase MTF and the SERVA enzymes [19]. For the Liberase MTF group, collagenase and thermolysin concentrations were adjusted separately and perfused into the pancreas simultaneously. The SERVA enzymes, however, were delivered separately – collagenase first then Neutral protease after initial digestion in the Ricordi chamber. They found that 53 % of Liberase MTF islet isolations compared with 33 % SERVA isolations were successful (they deemed this >400,000 islet equivalents). The quality of the islets was similar using both enzymes. More recently results of a study comparing Liberase HI, SERVA Collagenase NB1/Neutral Protease and Liberase MTF/Thermolysin indicated that the latter was superior to the others in terms of digestion efficacy (percentage of tissue digested by weight) and insulin secretion in response to glucose in vitro [20].

In mid-2008 another enzyme blend – VitaCyte collagenase HA – for clinical human islet isolation became available. Assessment of this blend showed that islet isolation outcomes were similar to those obtained with the SERVA enzymes [21]. This study demonstrated that the VitaCyte blend was more potent than the SERVA blend, and this did not cause deterioration of islet integrity, expressed as distribution of islet sizes, survival post-culture, insulin secretory capacity and cytokine expression.

A recent study evaluated three different enzyme combinations (ECs) to determine the optimal blend for isolating large numbers of high quality islets [22]. The ECs included the standard

SERVA NB1collagenase + NB Neutral protease (EC-A) and VitaCyte-CI*zyme*™ Collagenase-HA + VitaCyte – CI*zyme*™ Thermolysin (EC-F). These were compared to a new enzyme mixture (NEM) consisting of VitaCyte – CI*zyme*™ Collagenase HA + SERVA NB Neutral protease. The NEM consistently achieved higher islet yields from deceased donor pancreases ($p < 0.001$) than other standard ECs and met release criteria for transplantation from 8 of 10 consecutive pancreases, compared with 3 of 13 from EC-A and 7/19 from EC-F. All but one patient transplanted with islets isolated using the NEM exhibited adequate basal and stimulated C-peptide levels similar to patients in other enzyme groups.

A similar study compared the efficacy of SERVA NB1 collagenase with either a high activity-grade, low endotoxin level, neutral protease or thermolysin, in clinical islet isolation [23]. They reported that both combinations generated islets of a clinical grade. A retrospective analysis of SERVA NB1 collagenase and NB Neutral Protease with the NB1 and high activity neutral protease demonstrated there was no difference in islet mass or viability between the two groups or favourable 1 month post-transplant outcomes.

Thus it is still not clear which collagenase blends and/or combinations of such will provide the best outcomes for islet isolation, i.e. large numbers of high quality islets suitable for transplantation. In an attempt to shed light on what may be best, Rheinheimer et al. carried out a mixed treatment comparison (MTC) meta-analysis of studies that reported on human islet isolation and evaluated the effect of different enzyme blends on islet yield (IEQ/g pancreas), purity, viability and glucose-stimulated insulin release (SI) [24]. There were 755 articles retrieved from searches of Pubmed, Embase and Cochrane libraries. Of these, 15 were included in the MTC meta-analysis as they fulfilled the eligibility criteria. The analysed enzymes included Liberase HI, SERVA NB1, VitaCyte, Liberase MTF, Collagenase P (Boehringer Mannheim, Indianapolis, USA), Sevac (Crescent Chemical, Hauppauge, USA), Sigma V (Sigma, St. Louis, USA), Recombinant (Roche, Penzberg, Germany) and Collagenase Custom (Roche, Indianapolis, USA). This comprehensive analysis concluded that with regards to islet yield, purity and viability, the digestion enzymes currently being used for human islet isolation were of similar efficiency. In regards to glucose-stimulated insulin release, this was improved with SERVA NB1 and VitaCyte when compared to Liberase MTF.

The packaging of collagenases and proteases separately mean adjustment of enzyme ratios can be made to take into consideration donor characteristics, provide the ability to add each enzyme separately as proteases can accelerate digestion and combine different enzyme blends in order to improve islet yield. However, there is still difficulty in isolating sufficient numbers of high quality islets from young donors and marginal donors. In a small number of cases, our Unit perfused a reduced concentration of Liberase MTF + Thermolysin, as compared with our standard, into the pancreas of donors <25 years, with limited success (unpublished data). Another group, Shimoda et al., manually introduced a high concentration of collagenase NB1 + Neutral protease into pancreases from donors <30 years and compared islet isolation outcomes with donors of a similar age where a lower concentration of enzyme was automatically perfused into the organ [25]. They found the ratio of embedded islets was lower and the islet equivalents per pancreas weight, was higher in the trial group when compared to the standard group. This needs to be confirmed, as it was only performed in a small number of isolations.

Our Unit routinely now uses the SERVA GMP enzymes and reconstitutes Collagenase NB1 and Neutral protease NB during trimming and decontamination of the pancreas. These are combined immediately prior to infusion; the concentration can vary depending on the organ assessment but we use a range between 18 and 24 U/g pancreas of collagenase and 1–2 DMC U/g pancreas of Neutral protease. This range of concentration seems to work adequately in most cases. Varying the enzyme concentrations beyond these does not seem to alter our outcomes when we first started using the SERVA GMP enzymes.

7.1.3 Distension of the Pancreas with Enzyme

Islets are released from the human pancreas via infusion of digestive enzyme through the main pancreatic duct, which breaks off into smaller ducts that penetrate the exocrine portion of the pancreas. To ensure the release of a maximal number of islets, it is important that the enzyme is distributed evenly throughout the organ. This may be achieved through the cannulated pancreatic duct via two methods: hand distension or pumping with a mechanical perfusion device under pressure control.

Hand distension, has been used successfully by our islet isolation unit since the commencement of our clinical islet isolation programme in 2002 [9, 16]. After decontamination and trimming of the pancreas, it is cut in half and the duct on both halves is cannulated. The enzyme solution is then pumped through the cannulated duct using a 50 mL syringe and pulsatile motion, as can be seen in Fig. 7.2a. This allows the operator to monitor and adjust the enzyme delivery according to the physical attributes of the pancreas e.g. fibrous content or duct architecture. Figure 7.2b is an example of a pancreas optimally distended with enzyme solution following hand distension. Our isolation unit trialed hand held manometry to monitor the pressure while pumping enzyme solution into the pancreas, however, we found that it required extra time for set-up, was cumbersome with additional lines and connectors and did not enhance the outcome of the islet isolation procedure. Other groups load enzyme into the pancreas using a pump and a recirculating perfusion device system allowing enzyme infusion under controlled pressure. An early study comparing the two techniques showed that islet yield was higher post purification with the perfusion device but in vitro islet function as measured by glucose perifusion was no different between the two groups [26].

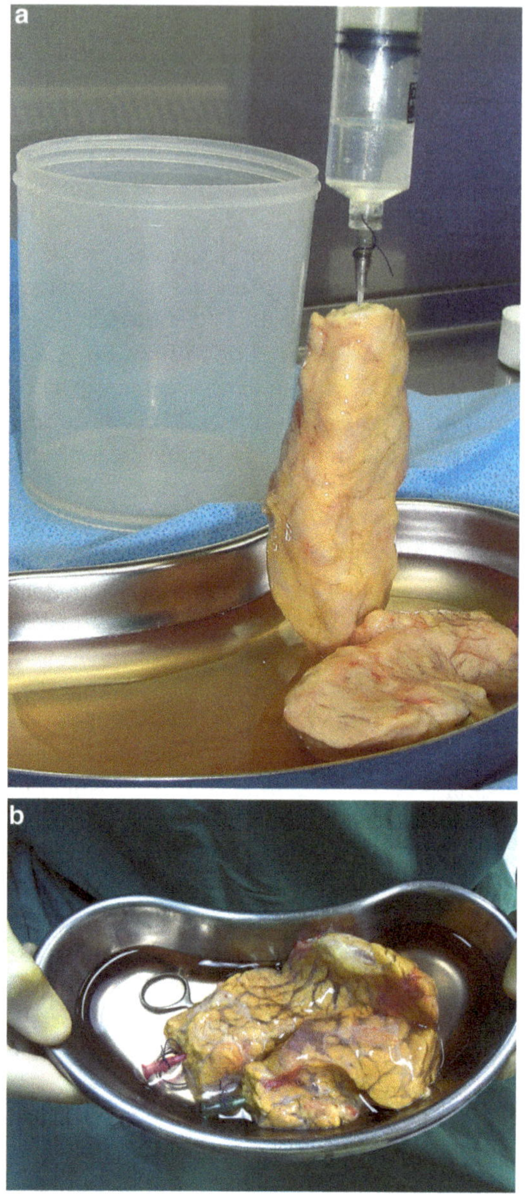

Fig. 7.2 (**a**) Is a photo of a pancreas being hand distended with collagenase using a 50 ml syringe. The collagenase enzyme is injected slowly in a pulsatile fashion via the cannula in the pancreatic duct. Note also that the pancreas has been dissected in half and the duct on both halves of the pancreas have been cannulated to allow for injection with complete and even distension of the gland, as can be seen in (**b**)

7.1.4 Release of the Islets Following Distension

Following the complete distension of the pancreas with enzyme solution, the organ is chopped into ~2 cm^3 pieces and placed into a metal dissociation chamber (Ricordi Chamber) containing stainless steel ball bearings. The separation of the islet cells from the acinar and connective tissue is achieved in the circulating system and a Ricordi chamber as seen in the digestion circuit consists of a closed circulation tubing system which circulates the collagenase and media containing the pancreas (Fig. 7.3a, b). The chamber and fluid are warmed to 37 °C and the pancreas is prevented from blocking the tubing by the use of a 500 μm mesh in the lid of the chamber. The chamber also contains sterile stainless steel or other type ball bearings that aid in the breaking up of the digesting pancreas tissues as the chamber is shaken gently. The islets pass through the mesh and continuous biopsy is used to identify the point at which the isolation has progressed to release of islets from the acinar and other tissues.

This stainless steel chamber called 'the Ricordi chamber' is connected to a tubing system is and was introduced by Ricordi and colleagues in 1988. It was termed the 'automated method' [4]. Enzyme solution is recirculated through the tubing system and the chamber is gently shaken to aid the enzymatic break down with mechanical disruption of the pancreatic tissue by the aid of the ball bearings.

The shaking of the chamber can be carried out manually i.e. by hand (Fig. 7.3a) or mechanically using a 'shaker'. A recent paper that compared manual with mechanical shaking reported that hand shaking yielded more islets with better integrity than mechanical shaking [13]. They found that digestion times were longer but yields higher and more pancreas digest collected, regardless of the enzyme used.

Our Unit uses hand shaking of the Ricordi chamber as it enables us to monitor the digestion process by sensing the disruption of the gland with the enzyme. It allows us to better control the shaking intensity depending on how dissociated the pancreas tissue feels e.g. is the tissue moving more freely in the chamber, is it pulling apart easily or are there still large intact pieces in the chamber. The agitation intensity is changed depending upon the amount of tissue disruption that has occurred. Shaking intensity is decreased as the gland pulls apart as observed as biopsies are collected during the digestion process as can be seen being collected from a 3 way tap in the closed tubing circuit (Fig. 7.3b) [27].

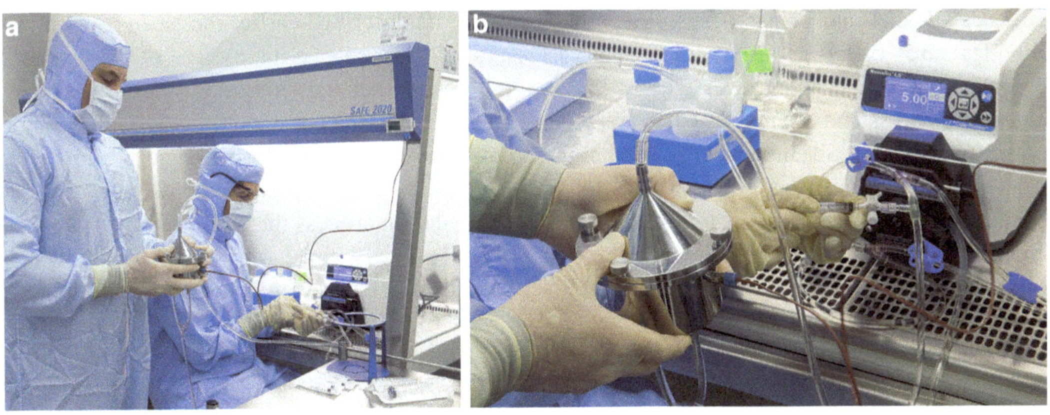

Fig. 7.3 (**a**) Shows the closed circulation tubing system with the Ricordi chamber in the foreground being gently shaken to aid in the breaking down of the pancreatic tissue. (**b**) Shaking of the chamber continues even whilst biopsies are collected for digestion assessment as can be seen being taken from the three-way tap in the circuit

7.1.5 Switching from Digestion to Collection

A critical step in the isolation of human islets for transplantation is determining when to switch from digestion of the pancreas to collection of the digested tissue. There is a fine line ensuring islets are adequately separated from the exocrine tissue, but have not been over-digested, resulting in fragmentation and destruction. It has been known for some time that cell-cell contact is necessary for islet function as disruption of the microanatomy alters insulin secretory responses [28]. Recently, Jaques et al. showed that the normal response to glucose was in part due to the engagement of the adhesion molecule E-cadherin between cells in contact with each other, which is Ca^{2+}-dependent [29]. So it is important to avoid excessive disruption during digestion of the pancreas. Further to this, using graph theory, Striegel et al., recently suggested that 'beta cell arrangement is dependent on its connectivity in order to maintain an optimal cluster size'. This must be kept in mind during digestion of the pancreas as disruption of this cell-cell contact may render the islets non-functional [30].

To achieve this fine balance, islet release is monitored constantly during digestion, by collecting biopsies, staining with dithizone (which stains zinc granules in the islet tissue) then viewing them under the microscope to review the presence of islet and acinar tissue (Fig. 7.4a). Biopsies are collected at regular intervals throughout the digestion process, early in the digestion process and when free islets are seen, collection begins. Collection involves switching from a closed system where enzyme solution is continually pumped through the Ricordi chamber and tubing system to an open system. In the open system, fresh solution, without enzyme, is pumped through the chamber and tubing system and digested tissue is collected into a cold media containing human albumin. Reduction in temperature means the enzyme cannot function optimally and dilution of the enzyme solution also inhibits its action. Figure 7.4b shows just fibrotic and ductal tissue remaining in the Ricordi chamber, demonstrating optimal digestion of the pancreas which indicates that the maximal number of islets have been released. Figure 7.4c shows a biopsy of the pancreas digest that has been sampled showing free islets stained red by dithizone stain in amongst acinar (yellow coloured cells) and other tissues.

The digest is then combined and washed twice in a cold wash media M199 containing human albumin, insulin and heparin to neutralize and further remove the digestive enzymes. It has been demonstrated that collagenase does not persist in the islets following washing during the isolation process [31] despite being detected immediately following infusion of collagenase through the ductal system [32], which eliminates concerns for patient safety. Following washing, the digest is placed in UW solution supplemented with a high concentration of human albumin, and heparin and insulin, for 30 min prior to purification. This 'quenching' step, allows uptake of starch into the acinar tissue altering its density and thus assisting separation during the next step of islet isolation – purification i.e. separation of the islets from the exocrine tissue.

7.2 Islet Purification

All islet isolation laboratories may well be different in shape, size and format but the basic principles of process and equipment remain the same. As described earlier the clean room plays an important role in the asepsis of the processing but another very important part of the overall process is the equipment in the facility. This equipment is essentially the same in most islet isolation laboratories as are the various stages in the isolation process that are made up of a number of very defined steps which have been described in significant detail in numerous studies [4, 33–35]. The development of the automated systems came following many early studies, which tried various methods for optimisation of the purification of the islets following the digestion phase. All revolved around optimisation of the already basic density separation with Ficoll-sodium diatrizoate, Dextran, Iodixanol and other radiological contrast media

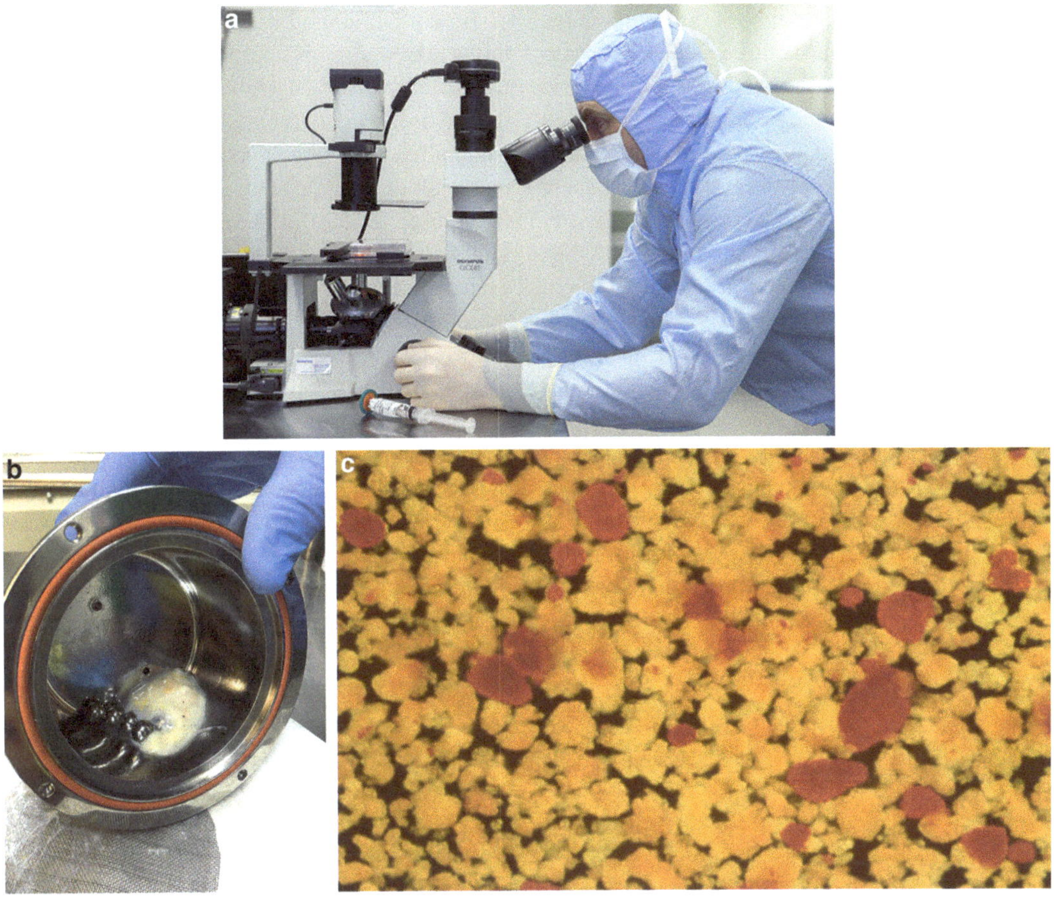

Fig. 7.4 (a) Examining biopsies of pancreatic digest stained with Dithizone to determine the appropriate time to switch from pancreas digestion to tissue collection and inactivation/removal of enzyme. (b) Remaining fibrous and ductal tissue in the Ricordi chamber following optimal digestion of the distended pancreas. (c) Is a biopsy from the pancreas digest that has been sampled showing free islets which are stained bright red with dithizone in amongst acinar and other tissues that are easily distinguished from the red staining islets

[35–40]. A number of variable mixtures of density gradients with agents such as the organ preservation solutions UW, Euro-Collins, to form variances of the Dextran and Ficoll gradients such as EuroFicoll or EuroDextran gradients all of which did not appear to have an advantage over the more traditional density gradients [36, 39, 40].

A number of studies also ran single density layer of Ficoll-sodium diatrizoate or Nycodenz at densities of 1.080 g/ml or 1.085 g/ml resulting in recovery of 47.4–77.4 % [37]. They suggested that further refinement of factors such as osmolality, viscosity, pH, ionic composition and temperature of iodinated density gradient media could provide continued improvement of islet purity and recovery.

There continued to be numerous studies over the years with ongoing changes and developments to improve purification outcomes. Chadwick et al., in 1994 described the technique of density-dependent purification of islets from several species of mammalian pancreata is improved by prior storage of the dispersed, collagenase-digested pancreas in suitable storage solutions, such as UW solution as cellular impermeants and colloids are important components [36]. In their study they dispersed tissues from 7 porcine and 7 human pancreata stored in UW or in solutions containing the impermeants lactobionate and raffinose, with either no added colloid or in the presence of the colloids hydroxyethyl starch, dextran 40, dextran 250, or Ficoll 400; hydroxyethyl starch-containing solutions in which the principal cation was sodium, rather than potassium, were also studied. Subsequent purification of islets on continuous linear density gradients of bovine serum albumin was then assessed by insulin/amylase assay of gradient fractions. Islet purity was slightly reduced using solutions containing impermeants but lacking a colloid, compared with using UW. In the combined presence of impermeants and a colloid, however, islet purity was similar to that obtained with UW, and for porcine pancreata, solutions containing Ficoll 400 or dextran 40 were slightly superior to UW. Purity was not, however, influenced by the sodium to potassium ratio of storage media. As such they concluded that impermeants and colloids are both essential components of solutions used to preserve pancreatic tissue before islet purification and specifically during collagenase digestion/density gradient purification [36].

The currently used and run density gradients have evolved from these earlier forms of density gradients. The current density separation media used in the major units around the world include; Ficoll, Dextran, Biocoll and various radiological contrast agents such as Iodixanol [16, 33, 34, 36, 41–47].

Ficoll or Ficoll-Paque density gradients are solutions of high molecular weight sucrose polymers and sodium diatrizoate. Ficoll density gradient media are excellent for isolating viable islet cells in high yield and purity. Dextran is a complex branched glucan (polysaccharides made of many glucose molecules) composed of chains of varying lengths (from 3 to 2,000 kDa) [36, 43, 45, 48]. A water-soluble high molecular weight glucose polymer (ranging between MW 1,000 and 40,000,000), Dextran is produced by the action of bacteria from the family *Lactobacillaceae* and certain other microorganisms on sucrose or glucose. This was one of the earliest used density gradients used by a number of units early on but less so these days [9, 36, 38, 40].

Biocoll separating solution is a polymer with a molecular weight of approximately 400,000 Da. Densities of up to 1.1 g/ml can be adjusted using this hydrophilic polymer. For optimal pH and osmolality, adjusting Biocoll with an acid, preferably amidotrizoeic acid, and sodium hydroxide is required. Biocoll with densities of 1.077 and 1.090 g/ml are already adjusted as commercially available separating media. In our unit we use the pre-prepared Biocoll gradients for islet purification [16].

Iodixanol is a radiologic contrast agent, sold under the trade name Visipaque; it is also sold as a density gradient under the name OptiPrep. It is the only iso-osmolar contrast agent, with an omolality of 290 mOsm/kg H_2O, the same as blood. It is sold in two main concentrations 270 and 320 mgI/ml – hence the name Visipaque 270 or 320 which are predominantly used by a number of the major units within Asia [44, 46, 47, 49].

The changes that have evolved in the density gradients have also been matched in advances in the equipment. One significant development that made advances to processing was the automated purification processing step using the IBM 2991 COBE cell separator as it reduced the time required for purification, shortening it to one fourth the usual time and total processing time to about half as long. Moreover, a team of fewer laboratory staff is now able to prepare islets for transplantation, significantly reducing overall costs [39, 50]. One of the earlier studies by Vargas et al. demonstrated major improvements to preparation of human islets for transplantation adopting the then use of the IBM 2991 COBE cell separator and a metrizamide/Ficoll density medium that was relatively easy to prepare. Using 27 pancreatic glands processed with the COBE cell separator, they showed a dramatic improvement of recovery and viability in these preparations when compared retrospectively with manual gradients. They concluded that the automatic cell separator and separation medium were major advances to the then islet purification methods [50]. Quite clearly the adoption of these techniques and the preference for the use of the automated technique utilising multiple density gradients on the COBE® 2991™ Cell Processor fast became the mainstay of the process which is still used the same today [35, 36, 38, 40, 41, 43].

The COBE® 2991™ Cell Processor which has been the main stay of the isolation process for many years involves some modifications to the stock cell separator. These were to develop cooling of the machine to ensure the density separation gradients that the islets were loaded onto remained cool, as did the islets. This is because a suitable hypothermic environment would prevent interaction with the toxic density gradients preventing ischaemic cell injury but also maintain the density of the temperature sensitive gradient solutions during the purification process. To this end, a number of modifications to many COBE® 2991™ Cell Processors were undertaken in various units around the world. Two different approaches of controlled cooling of the COBE® 2991™ cell separator for islet purification have been undertaken. The first method was to modify the machine itself and this was done by a number of methods that have included; water cooling, air cooling and even an electronically controlled liquid nitrogen injection system (Geneva COBE cooling system) [41]. The second way was the use of the "Clean Room Cold Room" maintained at 1–4 °C such as was established in a number of units in the USA such as at the University of Illinois, Chicago and San Francisco [41]. Both methods demonstrated similar temperature gradients from the beginning to the end of centrifugation both being around 7 °C. COBE cooling systems can easily be adapted to a COBE® 2991™ cell separator and are efficient in maintaining gradient solutions at a defined low temperature during centrifugation [41, 43]. Our own unit has modified several COBE® 2991™ Cell Processors with water cooling which was performed by our own in-house engineering department and can be seen in Fig. 7.5a.

Running the COBE® 2991™ Cell Processor to purify the more pure islet component from the acinar contaminated tissues is performed in a very steady process with the aid of chilled density mixer devices that allow for the continuous loading of two differing density gradients. Prior to uploading the mixing density gradients, the heaviest density gradient is loaded onto the bottom of the pre-cooled COBE bag, then the mixing gradients are top loaded over this base gradient. As can be seen in Fig. 7.5b the two differing gradients are slowly mixed whilst being loaded into the already primed and centrifuging COBE bag. Once the gradients are loaded, the chilled pancreas milieu is top loaded over the preloaded density gradients as can be seen in Fig. 7.5c. The COBE® 2991™ Cell Processor is then run for a predetermined time (usually several minutes) prior to the gradient/pancreas mix being pumped out into chilled tubes or flasks that have cold media and 10% human albumin. Keeping the temperature cool means the islets are less stressed from changes in temperature gradients and also the effects of the density gradient are less affected at the lower temperatures. The separated purified fractions are then immediately biopsied from each of the collection tubes, stained with dithizone and the fractions assessed for the number of islets in each fraction and also the purity of each fraction. The tubes containing the most pure fractions are then washed several times and recombined as are the less pure fractions. Care is taken to ensure the more impure fractions are kept separated from the more pure fractions to ensure that only the most pure fractions are combined for eventual transplantation. Once combined, these are sampled and assessed by dithizone staining. Dithizone is a stain that binds zinc ions found in β cells but not exocrine tissue for assessing islet numbers, mass and purity [51, 52]. Although it does not allow differentiation between endocrine cell types, it remains the quickest and most simple method of estimating islet cell numbers and purity to allow for determination of the success of the isolation and more importantly for determination of seeding density of the cells when placed into culture. The numbers are assessed at this stage to ensure that the pure islet aliquots are cultured separately from the more impure acinar bound cells. This will be discussed in far greater detail in the following sections.

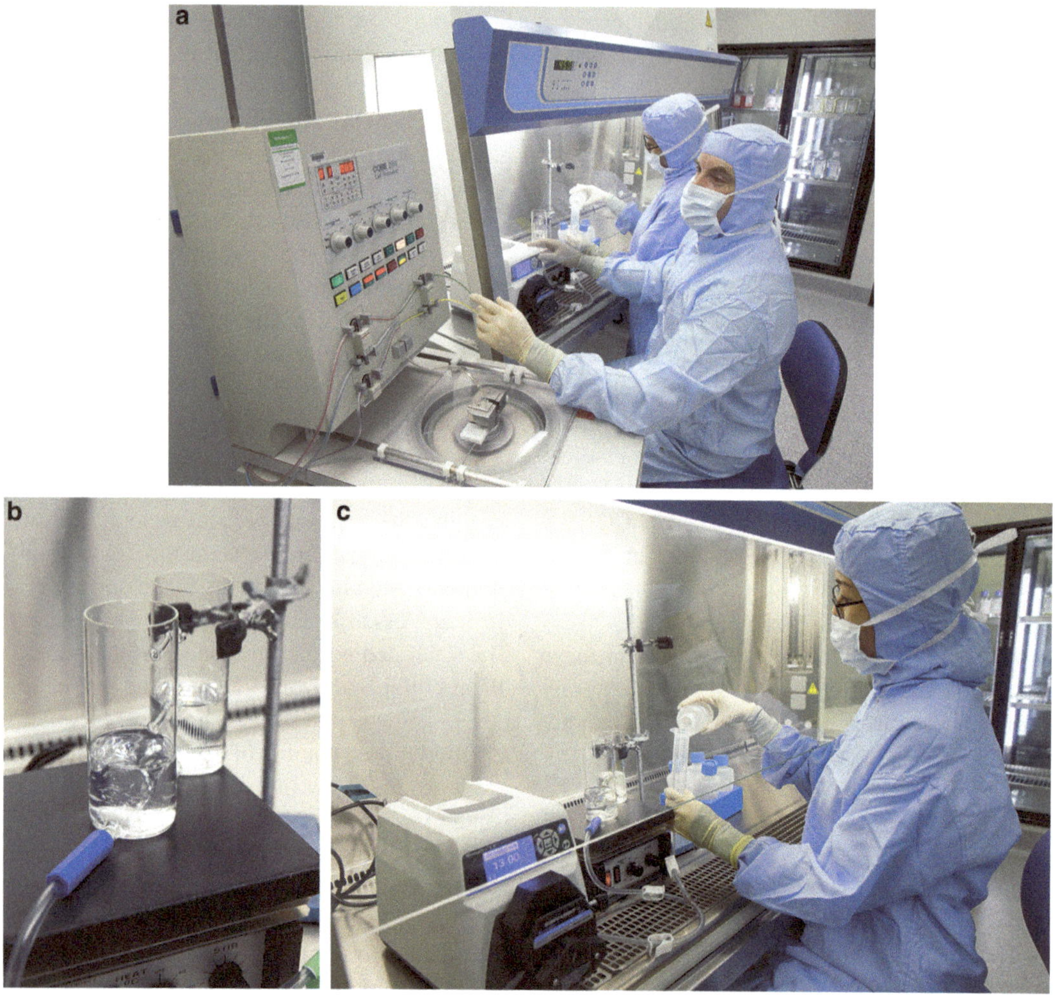

Fig. 7.5 (**a**) Laboratory setup for the islet purification phase. In the foreground the COBE® 2991™ Cell Processor has been set up with a COBE bag that is being loaded with a continuous density gradient by a gradient mixer as shown in (**b**). (**c**) Shows top loading of the density gradients with the pancreas digest in UW solution. This is then centrifuged on the COBE® 2991™ Cell Processor to allow for separation of the islet fractions from the remaining contaminating acinar and connective tissues

7.3 Islet Culture

7.3.1 To Culture or Not

Following isolation, human pancreatic islets may be transplanted immediately into a patient, or transferred into culture vessels and incubated for a specified duration before transplantation. Both approaches have been employed in the clinical setting with islets transplanted into patients within 2 h of isolation [9, 10, 53], and islets transplanted following up to 72 h of culture [16, 34, 54–57].

Various studies have been conducted with the aim of determining the cost and benefit of islet culture prior to transplantation. The main concern when culturing islets is the reduction in islet mass and functional capacity after the culture period [56, 58, 59]. Long-term culture studies (12–21 days) have shown a decrease in cell recovery rate, increase in DNA fragmentation, central necrosis and cell death, and loss of responsiveness to high glucose challenge [60–62]. However, in clinical application, islets are typically cultured no longer than 48 h before transplantation into the recipient.

Typical culture methods involve free-floating islets in non-adherent plastic culture flasks, an environment which does little to mimic the endogenous state where islets are supported within the extracellular matrix (ECM) [34, 63]. Destruction of the capillary networks surrounding the islets during the isolation process results in post-isolation hypoxic stress which contributes to islet loss in culture [63–65]. Studies have shown losses of up to 35 % cell mass following 72 h culture, with bioassays demonstrating that fresh islets allowed achievement of normoglycaemia with better glucose tolerance and stimulation indices compared to cultured islets [66]. Syngeneic transplant studies in rodents also demonstrate better outcomes with freshly isolated islets versus islets cultured for up to 1 week [67, 68].

However, other factors during the organ procurement and isolation process also play a role in loss of islets during culture. These include factors such as longer cold ischaemic time, lack of oxygen supplementation during organ preservation, larger islet size and lower preparation purity [56]. Management of these elements may be beneficial in improving the recovery rate of islets post-culture. For instance, separating islet preparations into fractions based on purity and culturing these in separate culture vessels may improve the recovery rate as enzymes released from dying tissue in the less pure fraction will not affect health cells in the fraction of higher purity [56].

It has also been suggested that those islets lost during culture are already determined at the point of isolation and the culture period serves to distinguish between these dying islets and to allow healthy islets of higher purity to be recovered for transplantation [69]. In addition, it allows time for performing quality assessment of the islet preparation such that any quality or contamination issues to be identified, preventing transplantation of poor preparations and improving transplantation outcomes [54]. The additional time afforded also allows for transport of islets to different transplant centres, as well as being able to have patients come to the transplant centre be health screened and commence on immunosuppression [54, 56, 70].

Kedinger et al. demonstrated prolonged survival of cultured human pancreatic islets transplanted into the liver of histo-incompatible patients with 70 % of recipients maintaining complete or partial glucose control up to 160 days post-transplant. This observation, along with similar studies involving culture of human islets of up to 7 days, suggests that short periods of in vitro culture are able to reduce the immunogenicity of islets [71–73].

An early study by Andersson et al. demonstrated that isolated human islets could be maintained in tissue culture for over 1 week without loss of alpha and beta cell function [74]. Other studies using porcine islets showed that although islet recovery gradually decreased as culture duration increased, the ability of recovered islets to reverse hyperglycaemia in mice improved with culture duration [75, 76]. Similar studies in porcine and human islets also generated the same outcome; improved islet function following culture in optimal media [58, 77]. This is supported by observations of higher ATP content of cultured islets in comparison to freshly isolated islets, suggesting recovery of islet metabolic and functional capacity while in culture [77, 78].

While most culture studies subject islets to long-term culture in the order of weeks to months [60–62], in general, clinical applications limit islet cell culture to the short-term – up to 72 h prior to transplant [34, 54–56]. Studies have shown that only a minimal loss of islet mass was seen in short-term cultures, with no significant changes in islet purity [56, 77]. It has also been suggested that decrease in islet mass may in part be due to islet recovery and reduction of swelling that occur during the perfusion process [69]. As transplant of large volumes of tissue are known to result in partial thrombosis of portal vein branches and changes to liver morphology [79–81], the reduction in packed cell volume after islet culture would correspondingly reduce the risk of portal pressure increase and thrombosis of the portal vein. However, increased expression of hypoxia/stress-related markers indicates that continued refinement of culture media and conditions have the potential to further improve islet recovery and maintenance of functional capacity post-culture [77, 82–84].

7.3.2 Base Culture Media

There are specific challenges faced in islet tissue culture as pancreatic islets consist of clusters of endocrine cells and are more akin to mini-organs as opposed to single cells. Changes to the media and techniques traditionally used are therefore required to provide optimal conditions for cell survival and recovery. As the ideal outcome of human pancreatic islet isolation is successful clinical transplantation of patients, the media used during culture must be defined, free of serum and xenoproteins, and contain little or no antibiotics or additional supplements.

Based on comprehensive testing of commercially available tissue culture medium, Connaught Medical Research Laboratories (CMRL) 1066 media was found to be optimal for islet cell culture [34, 58]. Originally used for culturing fibroblasts and kidney epithelial cells, it was found to be optimal for culturing human pancreatic islets [85], yielding cells with greater functional capacity compared to other media tested [58]. Several components of CMRL 1066 make it suitable for islet culture. One of these is glucose, essential for cellular ATP production, at a concentration of 5.5 mM, ideal for preserving responsiveness of human islets to glucose stimulation [86]. Prolonged exposure to higher glucose concentrations are known to lower insulin content and impair islet function [86].

Another form of islet culture media is defined Miami media, which consists of CMRL 1066 modified supplemented with nicotinamide, vitamin E and human serum albumin, in addition to insulin-transferrin-selenium, water-soluble linoleic acid, sodium pyruvate, zinc sulphate, HEPES and glutamine [87]. This modified form and similarly supplemented base media has also seen wide use in culture of human islets for the purposes of clinical transplantation [9, 16, 54, 57], although research is ongoing into the use of alternative media formulations [88, 89].

7.4 Additional Factors

7.4.1 Human Serum Albumin/Insulin-Transferrin-Selenium

The addition of serum or serum albumin to islet culture media has been comprehensively studied and these have shown that albumin possesses lipid-binding properties, therefore preventing complications associated with free lipid solubility and toxicity [90]. As xenoproteins must be absent in clinical transplant preparations, serum or serum albumin of human origin is added to culture media rather than fetal bovine serum (FBS; commonly used for culturing islets for research), although in vitro studies and bioassays have shown improved metabolic rate and function in islets cultured with fetal bovine serum [91].

A previous comparison between human serum (HS) and human serum albumin (HSA) determined that HSA was able to better preserve islet mass and secretory capacity compared to HS [92]. However, a more recent study observed increased viability, function and recovery rate of human islets cultured in media with HS added, although no significant difference in transplant outcome was seen in nude mouse bioassays [93].

The use of serum-supplemented versus serum-free media (SFM) has also been tested in long-term culture of human islets by comparing FBS-supplemented CMRL 1066, or CMRL supplemented with bovine albumin and insulin-transferrin-selenium (ITS) [61]. Insulin and insulin-like growth factors (IGF-1 and –II) have the potential to improve islet recovery and viability, and reduce mantle disintegration when added to islet culture media [61, 62]. The iron-binding properties of transferrin allow it to act as a stimulatory factor to islets, while selenium-dependent enzymes are able to scavenge reactive oxygen and nitrogen species, therefore preventing oxidative damage to cells in culture [90, 94].

Islets cultured up to 2 months in CMRL supplemented with ITS were found to have a higher recovery rate, viability and functional capacity compared to CMRL with FBS added [61]. A follow-up study using the same supplementation confirmed this as well as demonstrated improved islet function in vivo bioassays [95].

Current clinical programs utilise various concentrations of HSA as an additive to both culture media and the media used to transplant the islets into the patient. These are used at various concentrations but most clinical programs including our own utilise culture of islets in CMRL media a 10% HSA [16].

7.4.2 Other Various Additives

Free radical scavengers are frequently added to islet cell culture media to reduce oxidative damage. Two examples of such, nicotinamide (NIC) and desferrioxamine (DFO) have been found to protect against damage to chronic islet allografts and reduce the incidence of rejection [96–98]. In addition, neither NIC nor DFO were found to result in cell toxicity or impaired function when added to culture media [99, 100].

Another free radical scavenger, alpha-tocopherol (vitamin E) has also been shown to protect pancreatic islets against nitric oxide damage when added to islet culture media [101]. Supplementation with alpha-tocopherol during short-term culture was also found to moderately increase in vitro secretion of insulin of rat islets in response to glucose stimulation [102]. However, this effect was not maintained over culture periods longer than 3 weeks.

Activation of the free fatty acid (FFA) receptor GPR40 by linoleic acid was found to influence calcium channel signaling and result in increased glucagon secretion [103]. This FFA receptor has also been shown to play a role in free fatty acid-mediated insulin secretion [104]. Various studies have determined that accumulation of free fatty acids in pancreatic islets during glucose metabolism, assist in regulation of insulin secretion through voltage-dependent calcium channels [105, 106]. Additionally, addition of fatty acids to islet culture media was observed to increase cell proliferation and insulin secretion [107]. However, these beneficial effects were only observed with addition of polyunsaturated fatty acids such as arachidonic acid. Use of saturated fatty acids, such as palmitic acid, was found to increase production of reactive oxygen species, and impair islet viability and insulin secretion in response to glucose [107–109]. As such, care must be taken to select fatty acids suitable for supplementing culture media.

Pyruvate is a glycolytic intermediate and is transiently metabolized by islets, acting as a mitochondrial substrate [110]. In addition, pyruvate stimulates glucagon secretion and is known to improve recovery rate and glucose responsiveness when provided in islet culture media [110, 111].

Zinc is an important component of insulin synthesis, storage and secretion as proinsulin binds to zinc, forming a structure composed of six insulin molecules and two zinc atoms [112], and is stored as such in pancreatic beta cells [113]. Zinc also plays a signalling role, stimulating secretion of glucagon from alpha cells following secretion of insulin from beta cells [111]. The presence of zinc is therefore necessary to facilitate proper function of pancreatic islets. The addition of zinc chloride to culture media of rat islets demonstrated a reduction in apoptosis and mitochondrial oxidative stress [114]. Conversely, excessive zinc has been shown to decrease islet insulin content and negatively impact cell membrane integrity [115].

Glutamine, a non-essential amino acid, is frequently added to cell culture media due to its role as a cellular substrate source in addition to being an important precursor to glutathione, a free radical scavenger [116, 117]. Pig islets cultured in media supplemented with glutamine show increased cell recovery and viability, as well as improved stress resistance [118].

Perfusion of rat pancreata with glutamine during islet isolation was found to improve yield and viability of islets from pancreata subjected to 30 mins of warm ischaemia. The concentration of glutathione was also found to be increased with glutamine treatment [119]. A follow-up study in human pancreatic islet isolation observed similar results, with a reduction in percentage of apoptotic cells observed. Functional assessment in a nude mouse model showed that a higher percentage of mice achieved normoglycaemia within a shorter duration compared to mice transplanted with islets isolated without intraductal glutamine perfusion [120].

Treatment of rat islets with glutamine prior to transplantation was observed to protect against cytokine damage and nitric oxide-induced apoptosis through induction of heat shock protein (HSP) 70 and glutathione expression [121].

Glutamine also plays a role in regulating beta cell mass and function by stimulating biosynthesis and secretion of IGF-II, an autocrine ligand [122].

7.4.3 Culture Temperature

Various studies have attempted to determine the optimal temperature for islet cell culture. In general, islets are cultured in 5 % CO_2 at one of two temperatures: 22–24 °C (low) or 37 °C (high). Current clinical practice at most centres utilise culture at 37 °C for a period of 24 h and if further culture is required the temperature is reduced to 22–24 °C for no more than 12 h prior to transplantation [16, 123].

Human islets cultured at low temperatures have shown prolonged graft survival when transplanted into mouse models, suggesting that low temperature culture reduces islet immunogenicity [72, 73, 124]. In addition, low temperature culture lowers metabolic rate and is known to be beneficial for maintenance of islet morphology. Scharp et al. cultured human islets at 24 °C and showed reduced central necrosis and preservation of islet morphology after 1 week in culture [125]. A similar study in rat islets demonstrated that islets were able to maintain their morphology and function after 4 weeks in culture at 24 °C [126]. Both rat and pig islets showed decreased necrosis and cell death after low temperature culture, allowing higher recovery rates for cultured cells [60, 127].

Conversely, disintegration of the islet periphery was observed in hamster and pig islets cultured at low temperatures [60, 128]. Low temperature culture has also been shown to result in islet degranulation, impaired metabolism and glucose responsiveness [60, 129]. Escolar et al. showed partial inhibition of glucose-stimulated insulin secretion even at 27 °C, with complete inhibition at 17 °C [129]. Islets cultured at 24 °C demonstrated reduced insulin secretion in response to glucose after 1 week, although transferring those islets to 37 °C was found to be sufficient to return glucose responsiveness to initial levels [125, 126].

Culturing islets at 37 °C allows maintenance of islets close to physiological temperature. Pig islet studies show that islets cultured at 37 °C show increased viability, insulin content and secretion in response to glucose stimulation [60]. However, the higher metabolic rate is also known to result in a higher rate of necrotic cell death [126, 127, 129].

Currently, both temperatures are used for culture of human islets for clinical transplant with some groups culturing at 22 °C and others at 37 °C [9, 10, 16, 34]. Culture at 37 °C, followed by 22 °C has also used by some to take advantage of the benefits afforded by both culture temperatures [54, 55, 57]. A recent gene expression survey was conducted to compare porcine islets cultured for 6 days at either 37 °C or 22 °C, and for 5 days at 22 °C followed by 24 h at 37 °C [130]. This study demonstrated that islets cultured at 37 °C had marked reduction in expression of lymphocyte markers, lymphokines and chemokines. A smaller decrease in apoptotic and stress markers was also observed. Transfer of islets cultured at 22–37 °C for 24 h was sufficient to restore expression to the levels seen in the 6 day culture at 37 °C, suggesting that 37 °C may be an ideal temperature for islet cell culture [130]. However, care must be taken in extrapolating these results to culture of human islets as no similar gene expression studies have been conducted to identify the effect of culture temperature on human islets.

7.4.4 Seeding Density and Culture Vessels

Seeding density is another important consideration during islet culture as increasing islet

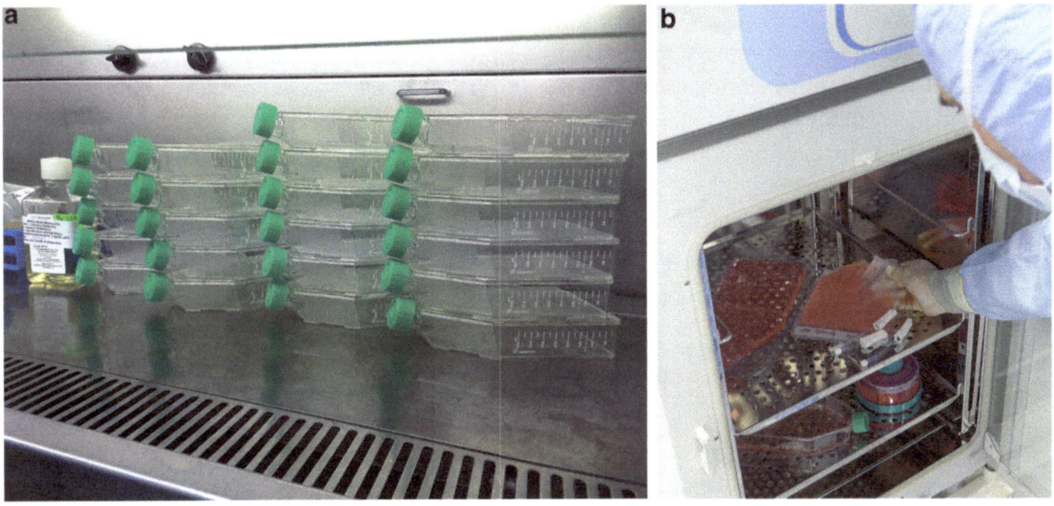

Fig. 7.6 (**a**) Multiple culture flasks are used for one pancreas worth of islets when cultured to ensure optimal cell density to try to reduce hypoxia and preserve cell viability. (**b**) Incubator inside the Human Applications Laboratory Clean Room showing examples of the various types of culture devices that can be used for culturing of islet cells. Some countries closely control the use of various flasks or vessels for the culturing of cells. Thus it is essential to ensure the appropriate standard device is used for culture of islets for clinical transplantation

density is thought to correlate to tissue necrosis [95]. A study by Brandhorst et al. showed that high islet densities result in a hypoxic environment, and a concomitant decrease in islet viability, insulin content and secretory function [131]. To avoid these adverse effects, isolated islets intended for clinical transplant are typically cultured at a seeding density at no more than 1,000 IEQ/ml culture media [34, 55].

In terms of culture vessels, islets are generally cultured in tissue culture flasks [9, 34, 57] or petri dishes [55, 56] (Fig. 7.6a). The amount of media used per vessel is generally kept low as the larger surface area and lower cell densities prevents hypoxia and preserves cell viability and function [132, 133].

Gas-permeable culture vessels have been trialed to improve oxygen delivery to islets in culture. A comparison of human islets cultured in vessels incorporating five different commercially available medical-grade gas-permeable membranes in addition to standard culture flasks identified an improved glucose stimulation response using one of the five tested [134]. Flasks with oxygen-permeable silicon rubber membranes have also been successfully used to culture human islets for 36–48 h at 10–20-fold higher densities than in conventional gas-impermeable flasks, with no significant loss in viability [135]. A distinct benefit of increasing cell culture density is that the number of culture flasks required is reduced, reducing the chances of contamination and islet loss during recombination for transplant.

In addition to flasks, a culture bag system has also been developed using a gas-permeable polyethylene membrane. Initial experiments using porcine islets demonstrated reduced cell loss, with improved viability and insulin secretion in response to glucose after 24 h culture, compared to islets cultured in conventional cell culture vessels [136]. This was also extended to human islet culture for transplantation; with good graft function observed in the two patients transplanted [136].

Human islets have also been cultured in a rotational cell culture system (RCCS) consisting of a high aspect ratio vessel (HARV) that is rotated horizontally throughout the culture period [137]. In this way, the islet cells are suspended in media, preventing settling and unwanted cell aggregation associated with necrosis while increasing oxygenation and maintaining cell-to-cell contact and signaling. A 10 day culture of human islets in a RCCS observed better preservation of islet morphology and function compared to static culture [137]. However, this culture system has not been tested as yet in the clinical setting. Our own unit has tested a number of different types of culture vessels for culturing islets these can be seen in (Fig. 7.6b) inside of an incubator within our isolation laboratory.

7.5 Quality Assessment Prior to Release for Transplantation

Various methods have been developed with the aim of assessing the quality of islet preparation following isolation and/or culture to ensure that the islet preparation will provide a safe and efficacious result. To be able to reach release criteria for transplantation the following are generally covered: cell mass, purity, viability, sterility and functional capacity. Quality assessment is particularly critical when characterizing islet preparations intended for clinical transplantation. It has previously been suggested that transplantation of poor quality islet preparations causes the inconsistencies observed in the ability of islet transplants to reverse diabetes [53], so quality control is essential to both determine the suitability of islets for transplant as well as to improve the chances of a long-term functional graft in recipients.

In the USA, the Food and Drug Administration regulations place islets isolated for transplant therapy under the biological products, and human cells and tissues, requiring the released preparation to demonstrate product stability and consistency between lots in addition to complying with standards of product identity, safety, purity and potency [138, 139].

However, due to donor variation and the nature of the isolation process, it may be difficult to apply regulations designed initially for therapeutics manufactured in a highly controlled situation to islet preparation release criteria wholesale. Other centres throughout the world typically employ methods intended to provide sensitive, accurate and reproducible quality assessments in the shortest time possible while ensuring release of a safe and efficacious preparation.

The release criteria formally accepted for the CITR program are based on islet count per recipient weight (5,000–20,000 IEQ/kg for the first transplant, and 3,000–20,000 for following transplants), with purity ≥30 %, viability ≥70 %, endotoxin concentration <5 endotoxin units (EU) per kg recipient weight, and no detectable organisms in a Gram stain prior to transplant, in addition to a glucose stimulation index (ratio of stimulated insulin secretion: basal insulin secretion) >1 [123, 140]. However, these criteria are generally accepted as guidelines to exclude poor preparations and may not necessarily relate to functional outcomes in vivo.

Criteria based on these are currently in formal use at the National Pancreas Transplant Unit at Westmead Hospital, Australia, where we require each islet preparation to be above determined thresholds of islet number/mass, viability, purity, and sterility before the product is released for transplantation (Table 7.1).

This section will briefly describe different indicators of islet quality most commonly assessed and used as release criteria, and their use in predicting islet survival and function in recipients. We will also describe a number of other useful methods that provide an indication of function but are NOT part of the formal release criteria.

7.5.1 Islet Cell Mass and Purity

Assessment of islet cell mass is a basic requirement for determining the suitability of an islet preparation for transplantation. As a proportion of islets may fail to engraft, apoptose or be destroyed by acute immune responses immediately following transplantation [141–143], a sufficient number of islets need to be transplanted to ameliorate this loss [10]. In Australia, for clinical islet transplantation this is currently a minimum of 4,000 islet equivalents (IEQ) per kilogram of recipient weight in order for the product to be released for transplantation along with also passing all other stipulated release criteria.

Table 7.1 Shows the release criteria used for product release for clinical islet transplantation at the National Islet Transplant Unit at Westmead Hospital

Parameter	Criteria limit	Result	Outcome
Islet Equivalents (IEQ) for transplantation	>200,000	(A–B)	Acceptable/not acceptable
A. Total IEQ post culture =			
B. IEQ taken for quality control =			
Islet equivalents/recipient body weight	>4,000/kg		Acceptable/not acceptable
Packed cell volume	<10 ml		Acceptable/not acceptable
Islet cell purity	>30 %		Acceptable/not acceptable
Islet cell viability	>70 %		Acceptable/not acceptable
Gram stain	Free of all organisms		No organisms/organisms present
Endotoxin assay	<25EU/50 ml (<0.5 EU/ml)	(C)	Acceptable/not acceptable
C. Total Endotoxin Units in sample (endotoxin value x sample volume)			
D. EU/kg Recipient body weight (C/recipient weight)	<5.0 EU/kg	(D)	Acceptable/Not acceptable

It has previously been shown in a pancreatectomized pig model that critical islet mass is essential for normalization of the glucose response after transplantation [144]. In human islet transplantation, sufficient islet mass is also known to be positively correlated to graft function, with various studies observing increased C-peptide and decreased requirement for insulin supplementation when larger numbers of islet cells were transplanted [42, 145].

However, studies of porcine and rat islet grafts showed that insulin content could be more closely correlated to the beta cell mass rather than just the number alone [146]. Microscopic analysis has indicated that up to 87 % of total islet volume in human islet preparations were found to be comprised of beta cells [147], although cytometric studies found this proportion to be closer to 60 %, suggesting some loss of beta cell mass during the isolation process [69]. Keymeulen et al. also observed that beta cell mass transplanted may be a better indicator of graft function post-transplant, rather than total islet cell mass. By controlling for the beta cell mass transplanted, they were able to standardise human islet transplant outcomes, with transplants of $>2 \times 10^6$ beta cells per kilogram recipient body weight found to maintain function up to 1 year post-transplant [148, 149].

Dithizone, a stain that binds zinc ions found in beta cells but not exocrine tissue, is commonly used to stain fresh and cultured islet tissue for assessing islet numbers, mass and purity [51, 52] as can be seen in (Fig. 7.7a) the biopsies are generally best viewed under an inverted microscope with a backlit stage. Although it does not allow differentiation between endocrine cell types, it remains the quickest and most simple method of estimating islet cell numbers and purity of the islet aliquot as seen in Fig. 7.7b where pure islets are stained vividly red with dithizone stain and some acinar tissue are attached to the islets and are not stained (yellow coloured tissues).

Briefly, islet numbers are counted after staining and divided into diameter categories with the smallest being 50 μm, and increasing incrementally by 50 μm. To standardise islet volume calculations, the islet equivalent (IEQ) was proposed as a measure of normalising islet volume based on the premise that 1 IEQ corresponds to a spherical islet of 150 μm diameter [52]. The number of islets in each diameter category is then converted using a pre-determined set of IEQ factors, and the sum of these calculated to obtain the total IEQ for each islet preparation [52] (Table 7.2).

Since then, several modifications have been proposed, such as adjustment of IEQ conversion factors to more accurately represent the

a b

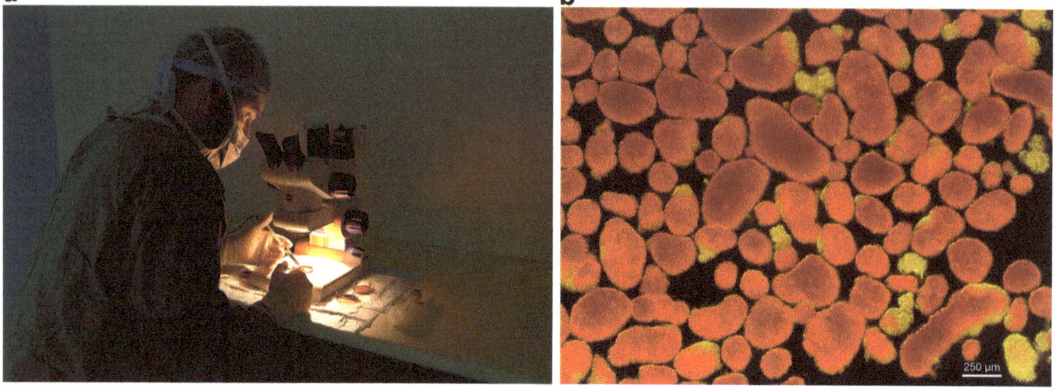

Fig. 7.7 (**a**) Counting and quantification of cell purity form an integral part of the quality assessment of the islet cells prior to release for transplantation or use in research. Cells are sampled in triplicate and placed into the wells of a six-well plate containing media and dithizone to stain the cells before examination under an inverted microscope. (**b**) A biopsy of human islets stained with dithizone which stains the islets red in contrast to any acinar or connective tissue that is not stained and is generally *yellow* when examined under an inverted microscope (Note that this biopsy is >90 % pure and has a good size distribution of islets)

proportion of different sizes of islets seen in a preparation [150], or to account for the fact that islets are not completely spherical with ellipsoid or irregular islets commonly observed [151, 152].

In addition to islet volume, islet size distribution is also known to have an impact on graft function with smaller islets demonstrating better function compared to large islets, both in rat models and human islet transplants [153, 154]. Small islets (diameter 50–150 μm) tended to be more viable, maintaining higher survival rates than large islets (diameter 150–300 μm) in both normoxic and hypoxic conditions [153, 154].

Table 7.2 To calculate the islet equivalent (IEQ), islet cells are categorised into different groups based on size increments of 50 μm, and the total number in each group is converted using a set of islet factors and combined to obtain the total IEQ

Islet diameter (μm)	Mean vol (μm^3)	IEQ: conversion into islets of 150 μm diameter
		Islet number (n) × islet factor
50–100	294,525	n/6.00
100–150	1,145,373	n/1.50
150–200	2,977,968	n × 1.7
200–250	6,185,010	n × 3.5
250–300	11,159,198	n × 6.3
300–350	18,293,231	n × 10.4
350–400	27,979,808	n × 15.8
Total IEQ		Sum of IEQ for each diameter category

Preparation purity is also an essential consideration in assessment of islets for transplant as acinar tissue remaining attached to islet cells post-isolation release proteases that contribute to islet cell death [56, 155]. Low purity islet preparations also demonstrate reduced viability and functional capacity [155–157]. Decreased purity and increased total preparation volume caused by attached acinar has been postulated to be a potential cause of thrombosis during infusion of the islet preparation into the portal vein during transplant as well as impair survival and engraftment of islets post-transplant [79, 158]. To assess islet purity post-isolation or -culture, islets are stained with dithizone to examine the proportion of cells remaining bound to acinar [157]. Islets may be accepted for release for clinical transplantation if the purity of the preparation is assessed to be over 30 % based on dithizone staining, with a total packed cell volume of less than 10 ml [9, 16].

7.5.2 Islet Cell Viability

Staining for islet cell mass alone is not sufficient to determine the quality of a preparation as islets can potentially be damaged and non-viable at the time of transplantation. As such if we transplant a large number of non-viable cells it provides a poor outcome in regards to function but also then provides an antigen load that can potentially sensitize the recipient to subsequent transplants. Obviously to transplant the best possible cells is of the utmost importance. To do this assessment of islet viability is an important factor in quality assessment and various methods are used to assess this. The currently accepted assay for islet viability involves staining with DNA-binding dyes to differentiate between live and dead cells based on membrane integrity, usually fluorescein diacetate (FDA) and propidium iodide (PI) [159, 160]. Based upon the CITR where they use FDA/PI and as performed in our own unit at Westmead a cut off of 70 % viability is a minimum for release of the product for transplantation [9, 16].

FDA is derived from fluorescein, a dye that fluoresces green. It diffuses passively across the cell membrane and is converted to fluorescein by esterase activity in the cytoplasm, causing live cells to fluoresce green under a 490 nm excitation wavelength [161]. Dead cells or dying cells are assumed to have minimal to no cytoplasmic esterase activity and therefore do not fluoresce green.

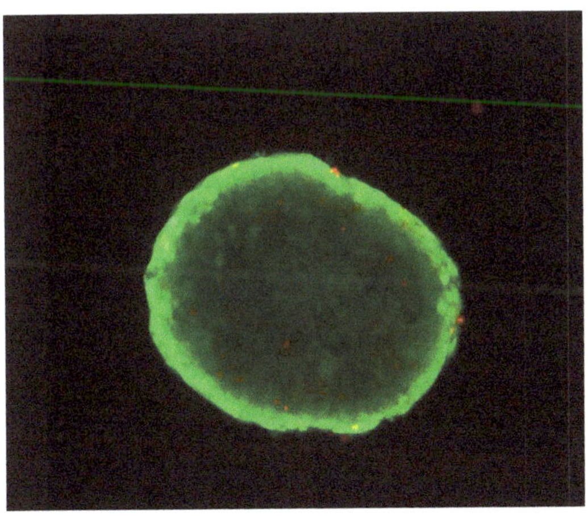

Fig. 7.8 Islet cells stained with FDA/PI showing live cells fluorescing *green* and damaged/dying cells fluorescing red

Counterstaining with PI (or a similar membrane-excluded dye such as ethidium bromide or ethidium homodimer-1 [162]) allows identification of damaged/dying cells exhibiting compromised membrane integrity as these will take up the stain, fluorescing red at 545 nm [159, 160] (Fig. 7.8). Obviously a threshold level of viable cells is required and according to our product release criteria, at least 70 % of cells must be viable before a preparation is deemed suitable for release for transplantation [9, 16].

However, use of FDA/PI can be subjective due to inconsistencies in dye concentration, incubation times, cell sample sizes or even imaging parameters. Membrane integrity is also assumed to indicate cell viability, although this may not necessarily be the case – islets judged as viable based on nucleic staining do not necessarily function and this has been shown in a number of studies including transplantation into mouse models, likely due to the fact that DNA-binding dye exclusion does not identify apoptotic cells [163].

Markers of apoptosis and necrosis have therefore been used in combination to allow a more accurate determination of islet viability. Ichii et al. have developed a method of simultaneously determining beta cell composition, viability and apoptotic cell percentage in a preparation using the zinc-binding dye Newport Green (NPG), apoptosis probe tetramethylrhodamine ethyl ester (TMRE) and membrane-impermeant 7-aminoactinomycin D (7-AAD) [69].

Zinc plays an essential role in insulin synthesis, storage and secretion in pancreatic beta cells [112, 113], and as NPG selectively binds zinc in an esterase-dependent fashion, viable beta cells can be identified using this marker [164]. Meanwhile, TMRE binds active mitochondria and decreased TMRE fluorescence serves as an indicator of cell apoptosis [165]. Finally, cells with membrane damage are stained with 7-AAD allowing identification of dead cells. By combining these dyes with high-throughput laser scanning cytometry and cytofluorimetry, a positive correlation was identified between viable beta cell mass and transplantation success in a mouse model [69]. This study also introduces the beta cell viability index based on the percentage of viable non-apoptotic beta cells as an indicator of graft survival and potential function post-transplant.

Our centre at Westmead Hospital, Australia conducts flow cytometric analysis on islet cells post-culture to determine this beta cell viability index, with indices of 0.5 or higher considered as satisfactory [9, 16]. However, this is not considered part of product release criteria as yet for transplantation and we have further studies ongoing to assess this for transplant release. In addition, although this method successfully allows characterisation of cells in an islet preparation, it also requires more time, a larger islet sample, technical expertise to both run and interpret the assay and the fluorescent cytometers equipped with lasers and filters suitable for sample analysis [166].

7.5.3 Sterility

As the main aim of clinical islet isolation is transplantation into a recipient, sterility of the final product is an essential criterion for product release. This is particularly important as recipients are immunosuppressed and thus are at an increased risk of infection should there be contaminants present in the final islet preparation [167]. Endotoxin contaminants are also known to contribute to islet cell damage and early graft loss, potentially due to direct binding of endotoxins to the CD14 receptor on pancreatic beta cells [168–170].

Microbial contamination potentially occurs at various stages throughout the islet isolation and culture process. Isolation and culture reagents are possible sources of endotoxins in islet preparations [168, 169], but the most likely source of contaminations is the donor duodenum during pancreas retrieval, as observed from testing of the solution in which the retrieved pancreas is preserved [171–173]. Scharp et al. observed that between microbial contamination was identified in up to 68 % of transport solutions processed each year [174]. The most common contaminant was identified as *Staphylococcus spp* [175, 176]. However, despite a high rate of contamination during retrieval, the majority of contaminants are removed during the isolation procedure, particularly during initial decontamination and purification processes [171, 173].

It is still essential to assess product sterility to determine the suitability of islet preparations for transplantation, and several measures are in place to reduce risk and assess the preparation after isolation and culture. Antibiotics (commonly ciprofloxacin) are added to culture media and aliquots are taken for Gram staining, endotoxin content assessment and microbiological culture both after isolation and pre-transplant after culturing [9, 172, 175]. In terms of product release, a negative Gram stain is required, in addition to endotoxin content under five endotoxin units (EU)/kg recipient weight [9, 172, 176]. A study encompassing over 358 islet isolations determined that all resulted in negative Gram stains and endotoxin levels under 5 EU/kg recipient weight [176]. Multiple studies have demonstrated that using these criteria, no clinical infection was observed in recipients and long-term graft survival remained unaffected [172, 176].

To culture for microbial sterility, sample aliquots are taken from media in which the donor pancreata are transported, media post-decontamination of the pancreas, after purification and post-culture. Two aliquots from each time point are inoculated aseptically into BACTEC™ culture vials (Becton Dickinson) containing broths specific for aerobic (tryptic soy broth) and anaerobic (soybean-casein digest broth) culture [9, 172, 175] (Figs. 7.9a, b).

In addition, samples are also cultured for fungi, mycoplasma and mycobacteria. However, assessment by culture is not used as release criteria due to the length of time required before results are obtained [172, 175]. In the event of a positive culture, appropriate antimicrobial prophylaxis is administered with little adverse effect on the recipient observed [172].

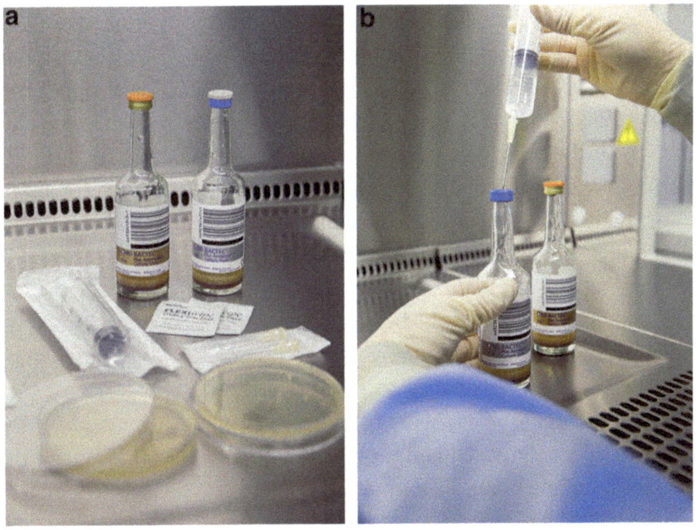

Fig. 7.9 (**a**) Shows the setup of equipment required for collection of the microbiology samples and (**b**) shows a sample of transport/organ perfusion fluid being collected from the receipt tray with the sample being sterilely inoculated into Bactec™ culture vials to be sent for culture and identification of any potential pathogens/contaminants

7.5.4 Useful Additional Tests; ATP/ ADP

Another method of determining viability is by measuring the amount of adenosine triphosphate (ATP) present. An early study by Brandhorst et al. observed that ATP levels in freshly isolated human islets were highly variable, and suggested a potential link between ATP content and graft efficacy as ATP is essential for cell homeostasis and function [78]. This has been demonstrated in a porcine-to-mouse islet transplantation model where ATP content was found to correlate positively with graft success [177].

However, while measuring ATP alone is able to provide an indication of cell viability, extending this to measurement of the ADP:ATP ratio by determining ATP before and after conversion of adenosine diphosphate (ADP) to ATP allows further differentiation between apoptotic (requires ATP) and necrotic (does not require ATP) cell death [178].

To measure ADP:ATP ratio in islet, a bioluminescent enzymatic assay was developed using synthetic firefly luciferases pyruvate kinase (PK) or pyruvate orthophosphate dikinase (PPDK), allowing assessment of islet ATP content and correlation of results to islet viability [179, 180]. Goto et al. were able to correlate islet ADP:ATP ratios to achievement of normoglycaemia in diabetic immune deficient mice transplanted with these cells [166]. As this assay can be performed with relative simplicity and speed, ADP:ATP ratio has been proposed as a viable method for quantitative measurement of islet energy status and functional capacity in determination of islet preparation suitability for transplantation. Despite this, it is not currently used as product release criteria for clinical transplantation of islets.

7.5.5 Oxygen Consumption Rate

Cell viability can also be assessed by measuring the mitochondrial oxygen consumption rate (OCR) as this is expected to correlate to the proportion of viable cells. An indication of fractional viability can then be obtained by normalizing this to cell DNA content (nmol/min.mg DNA).

Hellerstrom first developed a method for measurement of islet oxygen consumption in 1966 [181], and multiple methods have since been tested for assessing islet OCR. Sweet et al. employed a perifusion system to allow dynamic measurement of OCR in islets, while Papas et al. used a closed system involving continuous stirring for islet assessment [182, 183]. A different study measured islet OCR with an oxygen biosensor and fluorometric oxygen dyes in a culture plate system [184]. Using these methods, various groups were able to demonstrate correlation between oxygen consumption rate and the ability of islets to reverse diabetes in mouse models [182–187]. Pepper et al. further incorporated islet size assessment and showed that dividing the OCR value by the islet index allows accurate prediction of the ability of porcine islets to achieve normoglycaemia in diabetic nude mice [188].

In fact, it has been suggested that functional tissue mass (based on OCR assessment) is a better indicator of graft function as islets with high OCR measurements could be suitable for transplant at lower doses and vice versa. In this manner, Papas et al. were able to use variations in OCR/DNA measurements to adjust marginal mass of islets for transplantation, achieving successful outcomes in mouse models [189].

Islet OCR has also been measured in conjunction with glucose stimulation to determine both cell viability and functional capacity [190]. It has been demonstrated to be both indicative of transplant outcome as well as highly reproducible, making it a potential benchmark for islet quality assessment for transplantation [183, 186].

A high-throughput method for analysing islet oxygen consumption has also been developed using the extracellular flux analyser XF24 by Seahorse Bioscience (Billerica, MA) [191]. A specialised plate was designed to create a microenvironment within which islet bioenergetic status could be measured, including not only basal oxygen consumption, glucose-stimulated oxygen consumption, but also coupled and uncoupled respiration. While this assay requires specialised equipment and may take 5–7 h to conduct, it allows high-throughput and comprehensive analysis of the bioenergetic efficiency of the cells tested. However, at this point, while many centres do incorporate OCR assays for islet assessment, it is not currently used as formal criteria for product release in clinical transplantation.

7.5.6 Functional Analysis

Direct measurement of islet functional capacity in vitro has been proposed as another indicator of islet graft function after transplantation. This can be done by measuring islet insulin secretion after glucose stimulation, or by obtaining the stimulation index by comparing insulin levels before and after stimulation [4, 192]. Both static systems as well as dynamic perifusion assays have been developed for this purpose [52, 193]. Various studies have determined a range of around three- to fivefold increase in insulin secretion in response to glucose stimulation in vitro [194–196]. Unfortunately, comprehensive analyses of human islet preparations have determined that the glucose stimulation index does not reliably predict in vivo graft function and transplantation outcomes in mouse models [197, 198].

7.5.7 Mouse Bioassay

The gold standard for islet viability and function has generally been the ability of an islet preparation to reverse diabetes on transplantation into immunodeficient mice [52, 199, 200]. However, the only issue herein is that the mouse bioassay takes time to work (up to a week post-transplant) and as such cannot be used as part of the release criteria for clinical transplantation. As mentioned above, the various methods developed to assess islet quality (e.g. OCR, ADP:ATP ratio) are often judged by correlating assay results with achievement of normoglycaemia in vivo. To identify the ability of islets to reverse diabetes, a small number of islets are transplanted under the kidney capsule of athymic mice previously rendered diabetic using streptozotocin [52], following which blood sugar levels are monitored to determine achievement of normoglycaemia (Fig. 7.10a). Studies involving transplant of varying numbers of human, porcine or non-human primate islets into diabetic nude mice have determined that the higher the numbers of islets transplanted, the greater the chances of successful diabetes reversal [155].

However, islet viability and function could be adversely affected by additional time in culture if mouse transplants cannot be performed immediately [201]. In addition, studies have shown that when mice were transplanted with different islet preparations demonstrating similar values of viability, glucose response and endotoxin content, only 54 % were able to achieve normoglycaemia, indicating that additional factors influence transplant outcomes independently of islet quality

a b

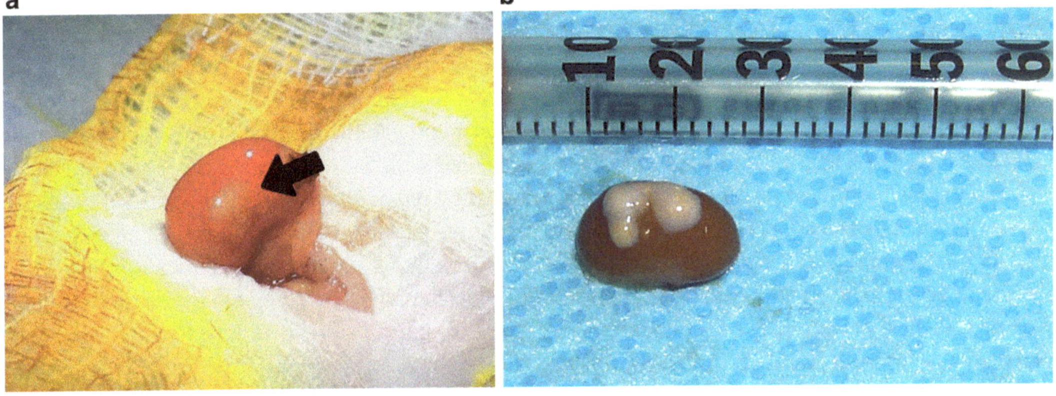

Fig. 7.10 Part of the quality assurance steps is the monitoring of the cells by the use of the mouse bioassay. (**a**) Shows human islets freshly transplanted under the kidney capsule of a mouse rendered diabetic by the use of streptozotocin. The *black arrow* is pointing to the transplanted islets under the kidney capsule. Blood sugar levels are monitored for a minimum of 1-month post-transplant. (**b**) Following long-term assessment the kidney with the islet graft is removed for macroscopic examination and histopathological assessment

[155]. These range from lower survival rates in mice with lower starting body weights, potential surgical complications, negative effects of streptozotocin induction of diabetes, as well as the length of time between induction and transplant [155, 202]. Rodents are also known to be less sensitive to porcine and human insulin, and transplanted islets are more susceptible to glucotoxicity in rodents immediately post-transplant, and as such may be less than ideal for assessing graft outcomes [203, 204]. In current clinical transplantation, success of diabetic reversal in mice following transplant of an islet aliquot is considered retrospectively following transplantation into the recipient [205]. Clear cut results with reversal of diabetes can take several days to occur and the grafts long-term function, macroscopic appearance and histopathology can only come many months following engraftment (Fig. 7.10b).

The only assessment criterion consistently found to correlate with in vivo islet graft function is transplanted mass [42, 140]. Currently, clinical islet transplant centres base the islet product release on islet yield (mass), islet viability, purity, endotoxin content and Gram stain results [9, 42, 52, 140, 172, 176]. The recommended release criteria follow these as a good guideline. However, with the advent of new technology and understanding on islet physiology, new methods are constantly being developed and refined to provide a prompt, reliable assessment of cell viability and function in islet preparations for clinical transplantation.

7.6 Bagging the Islets for Transplantation

The last stage of the overall rather complex process is the transplant procedure, which in itself is a variable process which relies on the success of the islet isolation process in the clean room to provide islets that are of an adequate number, viability, and free from any potential pathogens or contaminants.

The involved and extremely intricate series of steps to get to this point have ensured that the islets that have been prepared are of the highest quality and of sufficient numbers to provide a significantly beneficial outcome once transplanted into the recipient patient. The transplantation procedure is undertaken once all quality assurance steps have allowed the release of the islet product based upon the regulations of the Hospital's own institutional ethics committee, the local health authorities' regulations and ultimately the national or government regulatory body.

A significant outcome is to get to this point after undergoing the significant rigors of the isolation and quality assurance processes. But once achieved there are a number of potential options available to ensure good outcomes for the islet cells transplanted. The step immediately prior to transplantation is the bagging process to ensure for sterile and safe transport to the operating theatre or angiography theatre. The islets cells require to be deemed suitable for release from the isolation facility for clinical transplantation. To reach release criteria islet preparations are suggested to meet the following criteria as per Sect. 7.5 Quality Assessment Prior To Release For Transplantation; (1) Islet number of at least 4,000 IEQ/kg of recipient body weight, (2) packed islet tissue volume of less than 10 mL, (3) islet purity at least 30 %, 4) islet viability at least 70 % and (5) endotoxin level of less than 5 EU/kg recipient/h of infusion. In addition, the preparation has to be negative for microorganisms by Gram stain. Post-transplant assessment of the preparation should also include cultures for bacteria and fungus. If the preparation reaches these criteria and is accepted by the treating physician/surgeon, it is then bagged up for transplant. Islets are suspended in 100–150 ml of transplant grade CMRL 1066 media supplemented with 5–10 % HSA. The media and islets are loaded into transplant infusion bags immediately prior to being released and transported to the theatres. A second bag of infusion media/wash of 100–150 ml of transplant grade CMRL 1066 media supplemented with 5–10 % HSA is also loaded to allow the first bag with the islets to be washed to ensure that no islets are left in the bag or tubing when infused into the transplant recipient. Figure 7.11a shows human islets being loaded into a transplant bag minutes before it is taken to the operating theatres for transplant and Fig. 7.11b shows human islets in the transplant bag about to be placed into the transport container, which will be taken to the operating theatres for transplant as soon as possible.

7.7 Concluding Remarks

In this chapter we have outlined the many changes to and advances in the techniques for improving islet transplantation outcomes by improvements to islet isolation, culture and transplantation of clinical islets. However, islet transplantation still has limited application to the broader population of patients with T1D due to its reliance on the availability of cadaveric donor availability and

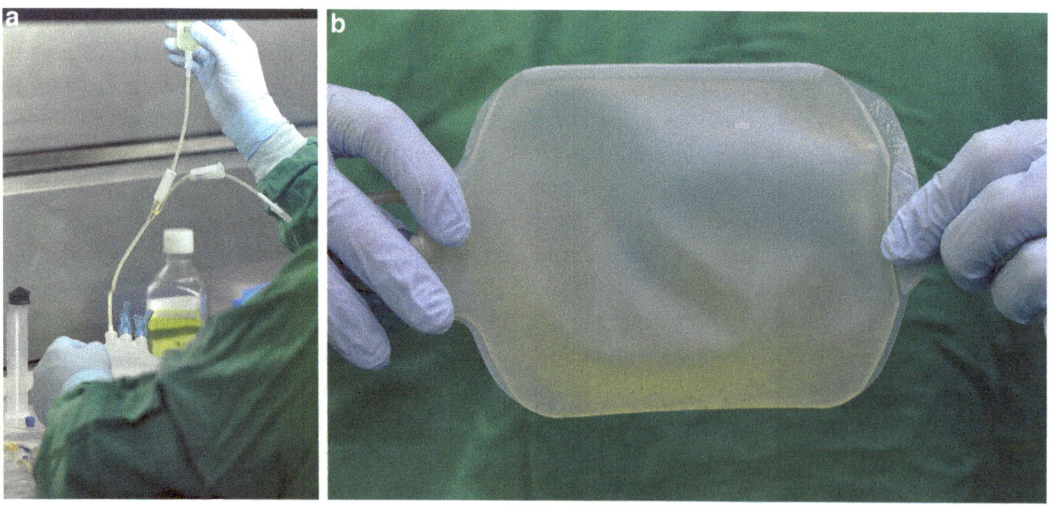

Fig. 7.11 The last step in the process following culture, quality assurance and final release of the islet preparation for transplant is the bagging for transplant. (a) Shows human islets being loaded into a transplant bag minutes before it is taken to the operating theatres for transplant. (**b**) Human islets seen as small white specs in the media in the transplant bag which is about to be placed into the transport container, which will be taken to the operating theatres or radiology suite for immediate transplant

selection, isolation results and transplant engraftment and as such we must strive to further improve these outcomes by further improving the processes involved in the isolation processes. Clearly great gains can be achieved by improvements to organ donation rates but ultimately the way in which can best improve our isolation outcomes is by improving the overall separation processes especially during digestion of the pancreatic tissue to protect the islets from the inherent hypoxic processes that they undergo whilst being extremely stressed in the process. Even changes to the way we culture and collect the islets from all steps in the processing can have an effect on the islets. With ongoing research in experimental and clinical studies, islet transplantation continues to be an accepted and very effective clinical treatment option to be able to offer patients suffering from type 1 diabetes with 'the prospect of shifting from a treatment for some to a cure for all' [206].

References

1. Hameed A, Yu T, Yuen L, Lam V, Ryan B, Allen R, et al. Use of the harmonic scalpel in cold phase recovery of the pancreas for transplantation: the westmead technique. Transpl Int. 2016;29:636–8.
2. Bockman DE. Anatomy of the pancreas. In: Go VLW, DiMagno EP, Gardner JD, Lebenthal E, Reber HA, Scheele GA, editors. The pancreas: biology, pathobiology, and disease. New York: Raven; 1993. p. 1–8.
3. Savari O, Zielinski MC, Wang X, Misawa R, Millis JM, Witkowski P, et al. Distinct function of the head region of human pancreas in the pathogenesis of diabetes. Islets. 2013;5(5):226–8.
4. Ricordi C, Lacy PE, Finke EH, Olack BJ, Scharp DW. Automated method for isolation of human pancreatic islets. Diabetes. 1988;37(4):413–20.
5. Kin T, Johnson PR, Shapiro AM, Lakey JR. Factors influencing the collagenase digestion phase of human islet isolation. Transplantation. 2007;83(1):7–12.
6. Linetsky E, Bottino R, Lehmann R, Alejandro R, Inverardi L, Ricordi C. Improved human islet isolation using a new enzyme blend, liberase. Diabetes. 1997;46(7):1120–3.
7. Barnett MJ, Zhai X, LeGatt DF, Cheng SB, Shapiro AM, Lakey JR. Quantitative assessment of collagenase blends for human islet isolation. Transplantation. 2005;80(6):723–8.
8. CITR Research Group. 2007 update on allogeneic islet transplantation from the Collaborative Islet Transplant Registry (CITR). Cell Transplant. 2009;18(7):753–67.
9. O'Connell PJ, Hawthorne WJ, Holmes-Walker DJ, Nankivell BJ, Gunton JE, Patel AT, et al. Clinical islet transplantation in type 1 diabetes mellitus: results of Australia's first trial. Med J Aust. 2006;184(5):221–5.
10. Shapiro AM, Lakey JR, Ryan EA, Korbutt GS, Toth E, Warnock GL, et al. Islet transplantation in seven patients with type 1 diabetes mellitus using a glucocorticoid-free immunosuppressive regimen. N Engl J Med. 2000;343(4):230–8.
11. ISCT. Risk of Bovine Spongiform Encephalopathy (BSE) in collagenase enzymes 2007. Available from: http://www.celltherapysociety.org/files/PDF/Resources/Risk_BSE_in_Collagenase_Enzymes.pdf.
12. Bucher P, Mathe Z, Bosco D, Andres A, Kurfuerst M, Ramsch-Gunther N, et al. Serva collagenase NB1: a new enzyme preparation for human islet isolation. Transplant Proc. 2004;36(4):1143–4.
13. Sabek OM, Cowan P, Fraga DW, Gaber AO. The effect of isolation methods and the use of different enzymes on islet yield and in vivo function. Cell Transplant. 2008;17(7):785–92.
14. Brandhorst H, Friberg A, Nilsson B, Andersson HH, Felldin M, Foss A, et al. Large-scale comparison of liberase HI and collagenase NB1 utilized for human islet isolation. Cell Transplant. 2010;19(1):3–8.
15. Barton FB, Rickels MR, Alejandro R, Hering BJ, Wease S, Naziruddin B, et al. Improvement in outcomes of clinical islet transplantation: 1999–2010. Diabetes Care. 2012;35(7):1436–45.
16. O'Connell PJ, Holmes-Walker DJ, Goodman D, Hawthorne WJ, Loudovaris T, Gunton JE, et al. Multicenter Australian trial of islet transplantation: improving accessibility and outcomes. Am J Transplant Off J Am Soc Transplant Am Soc Transplant Surg. 2013;13(7):1850–8.
17. Szot GL, Lee MR, Tavakol MM, Lang J, Dekovic F, Kerlan RK, et al. Successful clinical islet isolation using a GMP-manufactured collagenase and neutral protease. Transplantation. 2009;88(6):753–6.
18. Caballero-Corbalan J, Brandhorst H, Asif S, Korsgren O, Engelse M, de Koning E, et al. Mammalian tissue-free liberase: a new GMP-graded enzyme blend for human islet isolation. Transplantation. 2010;90(3):332–3.
19. O'Gorman D, Kin T, Imes S, Pawlick R, Senior P, Shapiro AM. Comparison of human islet isolation outcomes using a new mammalian tissue-free enzyme versus collagenase NB-1. Transplantation. 2010;90(3):255–9.
20. Qi M, Valiente L, McFadden B, Omori K, Bilbao S, Juan J, et al. The choice of enzyme for human pancreas digestion is a critical factor for increasing the success of Islet isolation. Transplant Direct. 2015;1(4).

21. Caballero-Corbalan J, Friberg AS, Brandhorst H, Nilsson B, Andersson HH, Felldin M, et al. Vitacyte collagenase HA: a novel enzyme blend for efficient human islet isolation. Transplantation. 2009;88(12): 1400–2.
22. Balamurugan AN, Loganathan G, Bellin MD, Wilhelm JJ, Harmon J, Anazawa T, et al. A new enzyme mixture to increase the yield and transplant rate of autologous and allogeneic human islet products. Transplantation. 2012;93(7):693–702.
23. O'Gorman D, Kin T, Pawlick R, Imes S, Senior PA, Shapiro AM. Clinical islet isolation outcomes with a highly purified neutral protease for pancreas dissociation. Islets. 2013;5(3):111–5.
24. Rheinheimer J, Ziegelmann PK, Carlessi R, Reck LR, Bauer AC, Leitao CB, et al. Different digestion enzymes used for human pancreatic islet isolation: a mixed treatment comparison (MTC) meta-analysis. Islets. 2014;6(4):e977118.
25. Shimoda M, Noguchi H, Naziruddin B, Fujita Y, Chujo D, Takita M, et al. Improved method of human islet isolation for young donors. Transplant Proc. 2010;42(6):2024–6.
26. Lakey JR, Warnock GL, Shapiro AM, Korbutt GS, Ao Z, Kneteman NM, et al. Intraductal collagenase delivery into the human pancreas using syringe loading or controlled perfusion. Cell Transplant. 1999;8(3):285–92.
27. Qi M, Barbaro B, Wang S, Wang Y, Hansen M, Oberholzer J. Human pancreatic islet isolation: part I: digestion and collection of pancreatic tissue. J Vis Exp. 2009;27:1125.
28. Hopcroft DW, Mason DR, Scott RS. Structure-function relationships in pancreatic islets: support for intraislet modulation of insulin secretion. Endocrinology. 1985;117(5):2073–80.
29. Jaques F, Jousset H, Tomas A, Prost AL, Wollheim CB, Irminger JC, et al. Dual effect of cell-cell contact disruption on cytosolic calcium and insulin secretion. Endocrinology. 2008;149(5):2494–505.
30. Striegel DA, Hara M, Periwal V. The beta cell in its cluster: stochastic graphs of beta cell connectivity in the Islets of Langerhans. PLoS Comput Biol. 2015;11(8):e1004423.
31. Cross SE, Hughes SJ, Clark A, Gray DW, Johnson PR. Collagenase does not persist in human islets following isolation. Cell Transplant. 2012;21(11): 2531–5.
32. Cross SE, Hughes SJ, Partridge CJ, Clark A, Gray DW, Johnson PR. Collagenase penetrates human pancreatic islets following standard intraductal administration. Transplantation. 2008;86(7): 907–11.
33. Matsumoto S, Noguchi H, Naziruddin B, Onaca N, Jackson A, Hatanaka N, Okitsu T, Kobayashi N, Klintmalm G, Levy M. Improvement of pancreatic islet cell isolation for transplantation. Proc (Baylor Univ Med Cent). 2007;20(4):357–62.
34. Qi M, Barbaro B, Wang S, Wang Y, Hansen M, Oberholzer J. Human pancreatic islet isolation: part II: purification and culture of human islets. J Vis Exp: JoVE. 2009(27).
35. Robertson GS, Chadwick D, Thirdborough S, Swift S, Davies J, James R, et al. Human islet isolation – a prospective randomized comparison of pancreatic vascular perfusion with hyperosmolar citrate or University of Wisconsin solution. Transplantation. 1993;56(3):550–3.
36. Chadwick DR, Robertson GS, Contractor HH, Rose S, Johnson PR, James RF, et al. Storage of pancreatic digest before islet purification. The influence of colloids and the sodium to potassium ratio in University of Wisconsin-based preservation solutions. Transplantation. 1994;58(1):99–104.
37. Hering BJ, Muench KP, Schelz J, Amelang D, Heitfeld M, Bretzel RG, et al. The evaluation of neutral density separation utilizing Ficoll-sodium diatrizoate and Nycodenz and centrifugal elutriation in the purification of bovine and canine islet preparations. Horm Metab Res Suppl. 1990;25:57–63.
38. Jindal RM, McShane P, Gray DW, Morris PJ. Isolation and purification of pancreatic islets by fluorescence activated islet sorter. Transplant Proc. 1994;26(2):653.
39. Lakey JRCT, Zieger MA. A prospective comparison of discontinuous EuroFicoll and EuroDextran gradients for islet purification. Cell Transplant. 1998;7(5): 479–87.
40. Weide LG, Damon-Burke M, Warkentin PI. Semiclosed system for human and porcine islet isolation using the COBE 2991 cell processor with the triple-bag processing sets. Transplant Proc. 1994;26(2):608–9.
41. Adewola A, Mage R, Hansen M, Barbaro B, Mendoza-Elias J, Harvat T, et al. Comparing cooling systems for the COBE 2991 cell separator used in the purification of human pancreatic islets of Langerhans. Cryo Lett. 2010;31(4):310–7.
42. Eckhard M, Brandhorst D, Winter D, Jaeger C, Jahr H, Bretzel RG, et al. The role of current product release criteria for identification of human islet preparations suitable for clinical transplantation. Transplant Proc. 2004;36(5):1528–31.
43. Friberg AS, Stahle M, Brandhorst H, Korsgren O, Brandhorst D. Human islet separation utilizing a closed automated purification system. Cell Transplant. 2008;17(12):1305–13.
44. Matsumoto S, Takita M, Chaussabel D, Noguchi H, Shimoda M, Sugimoto K, et al. Improving efficacy of clinical islet transplantation with iodixanol-based islet purification, thymoglobulin induction, and blockage of IL-1beta and TNF-alpha. Cell Transplant. 2011;20(10):1641–7.
45. Miki A, Ricordi C, Yamamoto T, Mita A, Barker S, Khan A, et al. Effect of human islet rescue gradient purification on islet yield and fractional Beta cell viability. Transplant Proc. 2008;40(2):360–1.
46. Noguchi H, Ikemoto T, Naziruddin B, Jackson A, Shimoda M, Fujita Y, et al. Iodixanol-controlled density gradient during islet purification improves

recovery rate in human islet isolation. Transplantation. 2009;87(11):1629–35.
47. Shimoda M, Itoh T, Sugimoto K, Takita M, Chujo D, Iwahashi S, et al. An effective method to release human islets from surrounding acinar cells with agitation in high osmolality solution. Transplant Proc. 2011;43(9):3161–6.
48. Eckhard M, Brandhorst D, Brandhorst H, Brendel MD, Bretzel RG. Optimization in osmolality and range of density of a continuous ficoll-sodium-diatrizoate gradient for isopycnic purification of isolated human islets. Transplant Proc. 2004;36(9):2849–54.
49. Mita A, Ricordi C, Messinger S, Miki A, Misawa R, Barker S, et al. Antiproinflammatory effects of iodixanol (OptiPrep)-based density gradient purification on human islet preparations. Cell Transplant. 2010;19(12):1537–46.
50. Vargas F, Vives-Pi M, Somoza N, Alcalde L, Armengol P, Martí M, Serradell L, Costa M, Fernandez-Llamazares J, Sanmartí A, Pujol-Borrell R. Advantages of using a cell separator and metrizamide gradients for human islet purification. Transplantation. 1996;61(11):1562–6.
51. Latif ZA, Noel J, Alejandro R. A simple method of staining fresh and cultured islets. Transplantation. 1988;45(4):827–30.
52. Ricordi C, Gray DW, Hering BJ, Kaufman DB, Warnock GL, Kneteman NM, et al. Islet isolation assessment in man and large animals. Acta Diabetol Lat. 1990;27(3):185–95.
53. Shapiro AM, Ricordi C, Hering BJ, Auchincloss H, Lindblad R, Robertson RP, et al. International trial of the Edmonton protocol for islet transplantation. N Engl J Med. 2006;355(13):1318–30.
54. Froud T, Ricordi C, Baidal DA, Hafiz MM, Ponte G, Cure P, et al. Islet transplantation in type 1 diabetes mellitus using cultured islets and steroid-free immunosuppression: Miami experience. Am J Transplant. 2005;5(8):2037–46.
55. Murdoch TB, McGhee-Wilson D, Shapiro AM, Lakey JR. Methods of human islet culture for transplantation. Cell Transplant. 2004;13(6):605–17.
56. Kin T, Senior P, O'Gorman D, Richer B, Salam A, Shapiro AM. Risk factors for islet loss during culture prior to transplantation. Transpl Int. 2008;21(11):1029–35.
57. Hering BJ, Kandaswamy R, Harmon JV, Ansite JD, Clemmings SM, Sakai T, et al. Transplantation of cultured islets from two-layer preserved pancreases in type 1 diabetes with anti-CD3 antibody. Am J Transplant. 2004;4(3):390–401.
58. Holmes MA, Clayton HA, Chadwick DR, Bell PR, London NJ, James RF. Functional studies of rat, porcine, and human pancreatic islets cultured in ten commercially available media. Transplantation. 1995;60(8):854–60.
59. Clayton HA, London NJ. Survival and function of islets during culture. Cell Transplant. 1996;5(1):1–12. discussion 3-7, 9.
60. Brandhorst D, Brandhorst H, Hering BJ, Bretzel RG. Long-term survival, morphology and in vitro function of isolated pig islets under different culture conditions. Transplantation. 1999;67(12):1533–41.
61. Fraga DW, Sabek O, Hathaway DK, Gaber AO. A comparison of media supplement methods for the extended culture of human islet tissue. Transplantation. 1998;65(8):1060–6.
62. Ilieva A, Yuan S, Wang RN, Agapitos D, Hill DJ, Rosenberg L. Pancreatic islet cell survival following islet isolation: the role of cellular interactions in the pancreas. J Endocrinol. 1999;161(3):357–64.
63. Wang RN, Rosenberg L. Maintenance of beta-cell function and survival following islet isolation requires re-establishment of the islet-matrix relationship. J Endocrinol. 1999;163(2):181–90.
64. Pileggi A, Ricordi C, Alessiani M, Inverardi L. Factors influencing Islet of Langerhans graft function and monitoring. Clin Chim Acta Int J Clin Chem. 2001;310(1):3–16.
65. Nagata NA, Inoue K, Tabata Y. Co-culture of extracellular matrix suppresses the cell death of rat pancreatic islets. J Biomater Sci Polym Ed. 2002;13(5):579–90.
66. Noguchi H, Naziruddin B, Jackson A, Shimoda M, Ikemoto T, Fujita Y, et al. Fresh islets are more effective for islet transplantation than cultured islets. Cell Transplant. 2012;21(2–3):517–23.
67. King A, Lock J, Xu G, Bonner-Weir S, Weir GC. Islet transplantation outcomes in mice are better with fresh islets and exendin-4 treatment. Diabetologia. 2005;48(10):2074–9.
68. Olsson R, Carlsson PO. Better vascular engraftment and function in pancreatic islets transplanted without prior culture. Diabetologia. 2005;48(3):469–76.
69. Ichii H, Inverardi L, Pileggi A, Molano RD, Cabrera O, Caicedo A, et al. A novel method for the assessment of cellular composition and beta-cell viability in human islet preparations. Am J Transplant. 2005;5(7):1635–45.
70. Ricordi C. Islet transplantation: a brave new world. Diabetes. 2003;52(7):1595–603.
71. Kedinger M, Haffen K, Grenier J, Eloy R. In vitro culture reduces immunogenicity of pancreatic endocrine islets. Nature. 1977;270(5639):736–8.
72. Ricordi C, Lacy PE, Sterbenz K, Davie JM. Low-temperature culture of human islets or in vivo treatment with L3T4 antibody produces a marked prolongation of islet human-to-mouse xenograft survival. Proc Natl Acad Sci U S A. 1987;84(22):8080–4.
73. Stein E, Mullen Y, Benhamou PY, Watt PC, Hober C, Watanabe Y, et al. Reduction in immunogenicity of human islets by 24 degrees C culture. Transplant Proc. 1994;26(2):755.
74. Andersson A, Borg H, Groth CG, Gunnarsson R, Hellerstrom C, Lundgren G, et al. Survival of isolated human islets of Langerhans maintained in tissue culture. J Clin Invest. 1976;57(5):1295–301.

75. Rijkelijkhuizen JK, van der Burg MP, Tons A, Terpstra OT, Bouwman E. Pretransplant culture selects for high-quality porcine islets. Pancreas. 2006;32(4):403–7.
76. Jimenez-Vera E, Davies S, Phillips P, O'Connell PJ, Hawthorne WJ. Long-term cultured neonatal islet cell clusters demonstrate better outcomes for reversal of diabetes: in vivo and molecular profiles. Xenotransplantation. 2015;22(2):114–23.
77. Ihm SH, Matsumoto I, Zhang HJ, Ansite JD, Hering BJ. Effect of short-term culture on functional and stress-related parameters in isolated human islets. Transpl Int. 2009;22(2):207–16.
78. Brandhorst D, Brandhorst H, Hering BJ, Federlin K, Bretzel RG. Large variability of the intracellular ATP content of human islets isolated from different donors. J Mol Med (Berlin, Germany). 1999;77(1):93–5.
79. Shapiro AM, Lakey JR, Rajotte RV, Warnock GL, Friedlich MS, Jewell LD, et al. Portal vein thrombosis after transplantation of partially purified pancreatic islets in a combined human liver/islet allograft. Transplantation. 1995;59(7):1060–3.
80. Maffi P, Angeli E, Bertuzzi F, Paties C, Socci C, Fedeli C, et al. Minimal focal steatosis of liver after islet transplantation in humans: a long-term study. Cell Transplant. 2005;14(10):727–33.
81. Bhargava R, Senior PA, Ackerman TE, Ryan EA, Paty BW, Lakey JR, et al. Prevalence of hepatic steatosis after islet transplantation and its relation to graft function. Diabetes. 2004;53(5):1311–7.
82. Maillard E, Juszczak MT, Clark A, Hughes SJ, Gray DR, Johnson PR. Perfluorodecalin-enriched fibrin matrix for human islet culture. Biomaterials. 2011;32(35):9282–9.
83. Moberg L, Johansson H, Lukinius A, Berne C, Foss A, Kallen R, et al. Production of tissue factor by pancreatic islet cells as a trigger of detrimental thrombotic reactions in clinical islet transplantation. Lancet (Lond Eng). 2002;360(9350):2039–45.
84. Stokes RA, Cheng K, Deters N, Lau SM, Hawthorne WJ, O'Connell PJ, et al. Hypoxia-inducible factor-1alpha (HIF-1alpha) potentiates beta-cell survival after islet transplantation of human and mouse islets. Cell Transplant. 2013;22(2):253–66.
85. Moore GE, Gerner RE, Franklin HA. Culture of normal human leukocytes. JAMA. 1967;199(8):519–24.
86. Eizirik DL, Korbutt GS, Hellerstrom C. Prolonged exposure of human pancreatic islets to high glucose concentrations in vitro impairs the beta-cell function. J Clin Invest. 1992;90(4):1263–8.
87. Al-Abdullah IH, Camillo R. Improvement of viability and function of pancreatic islets. Google Patents; 2003
88. Kerr-Conte J, Vandewalle B, Moerman E, Lukowiak B, Gmyr V, Arnalsteen L, et al. Upgrading pretransplant human islet culture technology requires human serum combined with media renewal. Transplantation. 2010;89(9):1154–60.
89. Kuhtreiber WM, Ho LT, Kamireddy A, Yacoub JA, Scharp DW. Islet isolation from human pancreas with extended cold ischemia time. Transplant Proc. 2010;42(6):2027–31.
90. Barnes D, Sato G. Methods for growth of cultured cells in serum-free medium. Anal Biochem. 1980;102(2):255–70.
91. Avgoustiniatos ES, Scott 3rd WE, Suszynski TM, Schuurman HJ, Nelson RA, Rozak PR, et al. Supplements in human islet culture: human serum albumin is inferior to fetal bovine serum. Cell Transplant. 2012;21(12):2805–14.
92. Lee RH, Carter J, Szot GL, Posselt A, Stock P. Human albumin preserves islet mass and function better than whole serum during pretransplantation islet culture. Transplant Proc. 2008;40(2):384–6.
93. Nacher M, Estil Les E, Garcia A, Nadal B, Pairo M, Garcia C, et al. Human serum versus human serum albumin supplementation in human islet pretransplantation culture. In vitro and in vivo assessment. Cell Transplant. 2015.
94. Brigelius-Flohe R, Banning A, Schnurr K. Selenium-dependent enzymes in endothelial cell function. Antioxid Redox Signal. 2003;5(2):205–15.
95. Gaber AO, Fraga DW, Callicutt CS, Gerling IC, Sabek OM, Kotb MY. Improved in vivo pancreatic islet function after prolonged in vitro islet culture. Transplantation. 2001;72(11):1730–6.
96. Bradley B, Prowse SJ, Bauling P, Lafferty KJ. Desferrioxamine treatment prevents chronic islet allograft damage. Diabetes. 1986;35(5):550–5.
97. Nomikos IN, Prowse SJ, Carotenuto P, Lafferty KJ. Combined treatment with nicotinamide and desferrioxamine prevents islet allograft destruction in NOD mice. Diabetes. 1986;35(11):1302–4.
98. Mendola J, Wright Jr JR, Lacy PE. Oxygen free-radical scavengers and immune destruction of murine islets in allograft rejection and multiple low-dose streptozocin-induced insulitis. Diabetes. 1989;38(3):379–85.
99. Eizirik DL, Sandler S, Welsh N, Bendtzen K, Hellerstrom C. Nicotinamide decreases nitric oxide production and partially protects human pancreatic islets against the suppressive effects of combinations of cytokines. Autoimmunity. 1994;19(3):193–8.
100. Langlois A, Bietiger W, Mandes K, Maillard E, Belcourt A, Pinget M, et al. Overexpression of vascular endothelial growth factor in vitro using deferoxamine: a new drug to increase islet vascularization during transplantation. Transplant Proc. 2008;40(2):473–6.
101. Burkart V, Gross-Eick A, Bellmann K, Radons J, Kolb H. Suppression of nitric oxide toxicity in islet cells by alpha-tocopherol. FEBS Lett. 1995;364(3):259–63.
102. Tajiri Y, Grill VE. Interactions between vitamin E and glucose on B-cell functions in the rat: an in vivo and in vitro study. Pancreas. 1999;18(3):274–81.
103. Wang L, Zhao Y, Gui B, Fu R, Ma F, Yu J, et al. Acute stimulation of glucagon secretion by linoleic

acid results from GPR40 activation and [Ca2+]i increase in pancreatic islet {alpha}-cells. J Endocrinol. 2011;210(2):173–9.
104. Schnell S, Schaefer M, Schofl C. Free fatty acids increase cytosolic free calcium and stimulate insulin secretion from beta-cells through activation of GPR40. Mol Cell Endocrinol. 2007;263(1–2): 173–80.
105. Wolf BA, Pasquale SM, Turk J. Free fatty acid accumulation in secretagogue-stimulated pancreatic islets and effects of arachidonate on depolarization-induced insulin secretion. Biochemistry. 1991;30(26):6372–9.
106. Turk J, Gross RW, Ramanadham S. Amplification of insulin secretion by lipid messengers. Diabetes. 1993;42(3):367–74.
107. Dixon G, Nolan J, McClenaghan NH, Flatt PR, Newsholme P. Arachidonic acid, palmitic acid and glucose are important for the modulation of clonal pancreatic beta-cell insulin secretion, growth and functional integrity. Clin Sci (Lond Engl 1979). 2004;106(2):191–9.
108. Zhou YP, Grill V. Long term exposure to fatty acids and ketones inhibits B-cell functions in human pancreatic islets of Langerhans. J Clin Endocrinol Metab. 1995;80(5):1584–90.
109. Lucena CF, Roma LP, Graciano MF, Veras K, Simoes D, Curi R, et al. Omega-3 supplementation improves pancreatic islet redox status: in vivo and in vitro studies. Pancreas. 2015;44(2):287–95.
110. Bottino R, Inverardi L, Valente U, Ricordi C. Serum-free medium and pyruvate improve survival and glucose responsiveness of islet beta cells in culture. Transplant Proc. 1997;29(4):1978.
111. Ishihara H, Maechler P, Gjinovci A, Herrera PL, Wollheim CB. Islet beta-cell secretion determines glucagon release from neighbouring alpha-cells. Nat Cell Biol. 2003;5(4):330–5.
112. Adams MJ, Blundell TL, Dodson EJ, Dodson GG, Vijayan M, Baker EN, et al. Structure of Rhombohedral 2 Zinc Insulin Crystals. Nature. 1969;224(5218):491–5.
113. Dodson G, Steiner D. The role of assembly in insulin's biosynthesis. Curr Opin Struct Biol. 1998;8(2):189–94.
114. Duprez J, Roma LP, Close AF, Jonas JC. Protective antioxidant and antiapoptotic effects of ZnCl2 in rat pancreatic islets cultured in low and high glucose concentrations. PLoS One. 2012;7(10):e46831.
115. Hoftiezer V, Berggren PO, Hellman B. Effects of zinc during culture of an insulin-producing rat cell line (RINm5F). Cancer Lett. 1985;29(1):15–22.
116. Amores-Sanchez MI, Medina MA. Glutamine, as a precursor of glutathione, and oxidative stress. Mol Genet Metab. 1999;67(2):100–5.
117. Wischmeyer PE, Musch MW, Madonna MB, Thisted R, Chang EB. Glutamine protects intestinal epithelial cells: role of inducible HSP70. Am J Physiol. 1997;272(4 Pt 1):G879–84.
118. Brandhorst H, Duan Y, Iken M, Bretzel RG, Brandhorst D. Effect of stable glutamine compounds on porcine islet culture. Transplant Proc. 2005;37(8):3519–20.
119. Avila JG, Tsujimura T, Oberholzer J, Churchill T, Salehi P, Shapiro AM, et al. Improvement of pancreatic islet isolation outcomes using glutamine perfusion during isolation procedure. Cell Transplant. 2003;12(8):877–81.
120. Avila J, Barbaro B, Gangemi A, Romagnoli T, Kuechle J, Hansen M, et al. Intra-ductal glutamine administration reduces oxidative injury during human pancreatic islet isolation. Am J Transplant. 2005;5(12):2830–7.
121. Jang HJ, Kwak JH, Cho EY, We YM, Lee YH, Kim SC, et al. Glutamine induces heat-shock protein-70 and glutathione expression and attenuates ischemic damage in rat islets. Transplant Proc. 2008;40(8): 2581–4.
122. Modi H, Cornu M, Thorens B. Glutamine stimulates biosynthesis and secretion of insulin-like growth factor 2 (IGF2), an autocrine regulator of beta cell mass and function. J Biol Chem. 2014;289(46):31972–82.
123. Balamurugan AN, Naziruddin B, Lockridge A, Tiwari M, Loganathan G, Takita M, et al. Islet product characteristics and factors related to successful human islet transplantation from the Collaborative Islet Transplant Registry (CITR) 1999–2010. Am J Transplant. 2014;14(11):2595–606.
124. Falqui L, Finke EH, Carel JC, Scharp DW, Lacy PE. Marked prolongation of human islet xenograft survival (human-to-mouse) by low-temperature culture and temporary immunosuppression with human and mouse anti-lymphocyte sera. Transplantation. 1991;51(6):1322–4.
125. Scharp DW, Lacy PE, Finke E, Olack B. Low-temperature culture of human islets isolated by the distention method and purified with ficoll or percoll gradients. Surgery. 1987;102(5):869–79.
126. Ono J, Lacy PE, Michael HE, Greider MH. Studies of the functional and morphologic status of islets maintained at 24 C for four weeks in vitro. Am J Pathol. 1979;97(3):489–503.
127. de Graaff MP, Wolters GH, van Schilfgaarde R. Endothelial cells in pancreatic islets and the effect of culture. Transplant Proc. 1994;26(3):1171.
128. Ilieva A, Yuan S, Wang R, Duguid WP, Rosenberg L. The structural integrity of the islet in vitro: the effect of incubation temperature. Pancreas. 1999;19(3):297–303.
129. Escolar JC, Hoo-Paris R, Castex C, Sutter BC. Effect of low temperatures on glucose-induced insulin secretion and glucose metabolism in isolated pancreatic islets of the rat. J Endocrinol. 1990;125(1): 45–51.
130. Mueller KR, Martins KV, Murtaugh MP, Schuurman HJ, Papas KK. Manufacturing porcine islets: culture at 22 degrees C has no advantage above culture at 37

degrees C: a gene expression evaluation. Xenotransplantation. 2013;20(6):418–28.
131. Brandhorst D, Brandhorst H, Mullooly N, Acreman S, Johnson PR. High seeding density induces local hypoxia and triggers a proinflammatory response in isolated human Islets. Cell Transplant. 2015.
132. Avgoustiniatos ES, Colton CK. Effect of external oxygen mass transfer resistances on viability of immunoisolated tissue. Ann N Y Acad Sci. 1997;831:145–67.
133. Lau J, Henriksnas J, Svensson J, Carlsson PO. Oxygenation of islets and its role in transplantation. Curr Opin Organ Transplant. 2009;14(6): 688–93.
134. Bentsi-Barnes K, Kandeel F, Al-Abdullah IH. Evaluation of human islet-specific functional quality cultured on different gas-permeable membranes. Transplant Proc. 2008;40(2):401–2.
135. Papas KK, Avgoustiniatos ES, Tempelman LA, Weir GC, Colton CK, Pisania A, et al. High-density culture of human islets on top of silicone rubber membranes. Transplant Proc. 2005;37(8):3412–4.
136. Goto M, Yoshikawa Y, Matsuo K, Shirasu A, Ogawa N, Takahashi H, et al. Optimization of a prominent oxygen-permeable device for pancreatic islets. Transplant Proc. 2008;40(2):411–2.
137. Murray HE, Paget MB, Downing R. Preservation of glucose responsiveness in human islets maintained in a rotational cell culture system. Mol Cell Endocrinol. 2005;238(1–2):39–49.
138. Linetsky E, Ricordi C. Regulatory challenges in manufacturing of pancreatic islets. Transplant Proc. 2008;40(2):424–6.
139. U.S. Department of Health and Human Services FDA. Guidance for industry: considerations for allogeneic pancreatic Islet cell products 2009 [updated 15/09/14. Available from: http://www.fda.gov/BiologicsBloodVaccines/GuidanceCompliance RegulatoryInformation/Guidances/Cellularand GeneTherapy/ucm182440.htm.
140. Papas KK, Suszynski TM, Colton CK. Islet assessment for transplantation. Curr Opin Organ Transplant. 2009;14(6):674–82.
141. Bennet W, Groth CG, Larsson R, Nilsson B, Korsgren O. Isolated human islets trigger an instant blood mediated inflammatory reaction: implications for intraportal islet transplantation as a treatment for patients with type 1 diabetes. Ups J Med Sci. 2000;105(2):125–33.
142. Bennet W, Sundberg B, Groth CG, Brendel MD, Brandhorst D, Brandhorst H, et al. Incompatibility between human blood and isolated islets of Langerhans: a finding with implications for clinical intraportal islet transplantation? Diabetes. 1999;48(10):1907–14.
143. Davalli AM, Scaglia L, Zangen DH, Hollister J, Bonner-Weir S, Weir GC. Vulnerability of islets in the immediate posttransplantation period. Dynamic changes in structure and function. Diabetes. 1996;45(9):1161–7.
144. Mellert J, Hering BJ, Liu X, Brandhorst D, Brandhorst H, Pfeffer F, et al. Critical islet mass for successful porcine islet autotransplantation. J Mol Med (Berlin Germany). 1999;77(1):126–9.
145. Ryan EA, Lakey JR, Rajotte RV, Korbutt GS, Kin T, Imes S, et al. Clinical outcomes and insulin secretion after islet transplantation with the Edmonton protocol. Diabetes. 2001;50(4):710–9.
146. Davalli AM, Ogawa Y, Scaglia L, Wu YJ, Hollister J, Bonner-Weir S, et al. Function, mass, and replication of porcine and rat islets transplanted into diabetic nude mice. Diabetes. 1995;44(1):104–11.
147. Pisania A, Weir GC, O'Neil JJ, Omer A, Tchipashvili V, Lei J, et al. Quantitative analysis of cell composition and purity of human pancreatic islet preparations. Lab Invest J Tech Methods Pathol. 2010;90(11):1661–75.
148. Keymeulen B, Gillard P, Mathieu C, Movahedi B, Maleux G, Delvaux G, et al. Correlation between beta cell mass and glycemic control in type 1 diabetic recipients of islet cell graft. Proc Natl Acad Sci U S A. 2006;103(46):17444–9.
149. Keymeulen B, Ling Z, Gorus FK, Delvaux G, Bouwens L, Grupping A, et al. Implantation of standardized beta-cell grafts in a liver segment of IDDM patients: graft and recipients characteristics in two cases of insulin-independence under maintenance immunosuppression for prior kidney graft. Diabetologia. 1998;41(4):452–9.
150. Buchwald P, Wang X, Khan A, Bernal A, Fraker C, Inverardi L, et al. Quantitative assessment of islet cell products: estimating the accuracy of the existing protocol and accounting for islet size distribution. Cell Transplant. 2009;18(10):1223–35.
151. Huang HH, Ramachandran K, Stehno-Bittel L. A replacement for islet equivalents with improved reliability and validity. Acta Diabetol. 2013;50(5):687–96.
152. Ramachandran K, Huang HH, Stehno-Bittel L. A simple method to replace islet equivalents for volume quantification of human islets. Cell Transplant. 2015;24(7):1183–94.
153. Lehmann R, Zuellig RA, Kugelmeier P, Baenninger PB, Moritz W, Perren A, et al. Superiority of small islets in human islet transplantation. Diabetes. 2007;56(3):594–603.
154. MacGregor RR, Williams SJ, Tong PY, Kover K, Moore WV, Stehno-Bittel L. Small rat islets are superior to large islets in in vitro function and in transplantation outcomes. Am J Physiol Endocrinol Metab. 2006;290(5):E771–9.
155. Loganathan G, Graham ML, Radosevich DM, Soltani SM, Tiwari M, Anazawa T, et al. Factors affecting transplant outcomes in diabetic nude mice receiving human, porcine, and nonhuman primate islets: analysis of 335 transplantations. Transplantation. 2013;95(12):1439–47.
156. Migliavacca B, Nano R, Antonioli B, Marzorati S, Davalli AM, Di Carlo V, et al. Identification of in vitro parameters predictive of graft function: a

study in an animal model of islet transplantation. Transplant Proc. 2004;36(3):612–3.

157. Kitzmann JP, Karatzas T, Mueller KR, Avgoustiniatos ES, Gruessner AC, Balamurugan AN, et al. Islet preparation purity is overestimated, and less pure fractions have lower post-culture viability before clinical allotransplantation. Transplant Proc. 2014;46(6):1953–5.
158. Gray DW, Sutton R, McShane P, Peters M, Morris PJ. Exocrine contamination impairs implantation of pancreatic islets transplanted beneath the kidney capsule. J Surg Res. 1988;45(5):432–42.
159. London NJ, Contractor H, Lake SP, Aucott GC, Bell PR, James RF. A microfluorometric viability assay for isolated human and rat islets of Langerhans. Diabetes Res (Edinburgh, Scotland). 1989;12(3): 141–9.
160. London NJ, Contractor H, Lake SP, Aucott GC, Bell PR, James RF. A fluorometric viability assay for single human and rat islets. Horm Metab Res Suppl Ser. 1990;25:82–7.
161. Kramer DN, Guilbault GG. A substrate for the fluorometric determination of lipase activity. Anal Chem. 1963;35(4):588–9.
162. Barnett MJ, McGhee-Wilson D, Shapiro AM, Lakey JR. Variation in human islet viability based on different membrane integrity stains. Cell Transplant. 2004;13(5):481–8.
163. Boyd V, Cholewa OM, Papas KK. Limitations in the use of Fluorescein Diacetate/Propidium Iodide (FDA/PI) and cell permeable nucleic acid stains for viability measurements of isolated islets of langerhans. Curr Trends Biotechnol Pharm. 2008;2(2):66–84.
164. Lukowiak B, Vandewalle B, Riachy R, Kerr-Conte J, Gmyr V, Belaich S, et al. Identification and purification of functional human beta-cells by a new specific zinc-fluorescent probe. J Histochem Cytochem. 2001;49(4):519–28.
165. Scaduto Jr RC, Grotyohann LW. Measurement of mitochondrial membrane potential using fluorescent rhodamine derivatives. Biophys J. 1999;76(1 Pt 1): 469–77.
166. Goto M, Holgersson J, Kumagai-Braesch M, Korsgren O. The ADP/ATP ratio: a novel predictive assay for quality assessment of isolated pancreatic islets. Am J Transplant. 2006;6(10):2483–7.
167. Taylor GD, Kirkland T, Lakey J, Rajotte R, Warnock GL. Bacteremia due to transplantation of contaminated cryopreserved pancreatic islets. Cell Transplant. 1994;3(1):103–6.
168. Berney T, Molano RD, Cattan P, Pileggi A, Vizzardelli C, Oliver R, et al. Endotoxin-mediated delayed islet graft function is associated with increased intra-islet cytokine production and islet cell apoptosis. Transplantation. 2001;71(1):125–32.
169. Vargas F, Vives-Pi M, Somoza N, Armengol P, Alcalde L, Marti M, et al. Endotoxin contamination may be responsible for the unexplained failure of human pancreatic islet transplantation. Transplantation. 1998;65(5):722–7.
170. Vives-Pi M, Somoza N, FernÁNdez-Alvarez J, Vargas F, Caro P, Alba A, et al. Evidence of expression of endotoxin receptors CD14, toll-like receptors TLR4 and TLR2 and associated molecule MD-2 and of sensitivity to endotoxin (LPS) in islet beta cells. Clin Exp Immunol. 2003;133(2):208–18.
171. Bucher P, Mathe Z, Bosco D, Oberholzer J, Toso C, Andres A, et al. Microbial surveillance during human pancreatic islet isolation. Transplant Proc. 2004;36(4):1147–8.
172. Kin T, Rosichuk S, Shapiro AM, Lakey JR. Detection of microbial contamination during human islet isolation. Cell Transplant. 2007;16(1):9–13.
173. Lakey JR, Rajotte RV, Warnock GL. Microbial surveillance of human islet isolation, in vitro culture, and cryopreservation. Clin Invest Med. 1995;18(3):168–76.
174. Scharp DW, Lacy PE, McLear M, Longwith J, Olack B. The bioburden of 590 consecutive human pancreata for islet transplant research. Transplant Proc. 1992;24(3):974–5.
175. Carroll PB, Ricordi C, Fontes P, Rilo HR, Phipps J, Tzakis AG, et al. Microbiologic surveillance as part of human islet transplantation: results of the first 26 patients. Transplant Proc. 1992;24(6):2798–9.
176. Gala-Lopez B, Kin T, O'Gorman D, Pepper AR, Senior P, Humar A, et al. Microbial contamination of clinical islet transplant preparations is associated with very low risk of infection. Diabetes Technol Ther. 2013;15(4):323–7.
177. Kim JH, Park SG, Lee HN, Lee YY, Park HS, Kim HI, et al. ATP measurement predicts porcine islet transplantation outcome in nude mice. Transplantation. 2009;87(2):166–9.
178. Bradbury DA, Simmons TD, Slater KJ, Crouch SP. Measurement of the ADP:ATP ratio in human leukaemic cell lines can be used as an indicator of cell viability, necrosis and apoptosis. J Immunol Methods. 2000;240(1–2):79–92.
179. Ishii S, Saito T, Ise K, Sato Y, Tsutiya T, Kenjo A, et al. Evaluation of energy state of islet independent of size using a newly developed ATP bioluminescence assay. Transplant Proc. 2005;37(8):3499–500.
180. Ishii S, Sato Y, Terashima M, Saito T, Suzuki S, Murakami S, et al. A novel method for determination of ATP, ADP, and AMP contents of a single pancreatic islet before transplantation. Transplant Proc. 2004;36(4):1191–3.
181. Hellerstrom C. Oxygen consumption of isolated pancreatic islets of mice studied with the cartesian-diver micro-gasometer. Biochem J. 1966;98(1):7c–9c.
182. Papas KK, Colton CK, Nelson RA, Rozak PR, Avgoustiniatos ES, Scott 3rd WE, et al. Human islet oxygen consumption rate and DNA measurements predict diabetes reversal in nude mice. Am

J Transplant Off J Am Soc Transplant Am Soc Transplant Surg. 2007;7(3):707–13.
183. Sweet IR, Gilbert M, Jensen R, Sabek O, Fraga DW, Gaber AO, et al. Glucose stimulation of cytochrome C reduction and oxygen consumption as assessment of human islet quality. Transplantation. 2005;80(8):1003–11.
184. Fraker C, Timmins MR, Guarino RD, Haaland PD, Ichii H, Molano D, et al. The use of the BD oxygen biosensor system to assess isolated human islets of langerhans: oxygen consumption as a potential measure of islet potency. Cell Transplant. 2006;15(8–9):745–58.
185. Papas KK, Bellin MD, Sutherland DE, Suszynski TM, Kitzmann JP, Avgoustiniatos ES, et al. Islet Oxygen Consumption Rate (OCR) dose predicts insulin independence in clinical islet autotransplantation. PLoS One. 2015;10(8):e0134428.
186. Sweet IR, Gilbert M, Scott S, Todorov I, Jensen R, Nair I, et al. Glucose-stimulated increment in oxygen consumption rate as a standardized test of human islet quality. Am J Transplant Off J Am Soc Transplant Am Soc Transplant Surg. 2008;8(1):183–92.
187. Sweet IR, Khalil G, Wallen AR, Steedman M, Schenkman KA, Reems JA, et al. Continuous measurement of oxygen consumption by pancreatic islets. Diabetes Technol Ther. 2002;4(5):661–72.
188. Pepper AR, Hasilo CP, Melling CW, Mazzuca DM, Vilk G, Zou G, et al. The islet size to oxygen consumption ratio reliably predicts reversal of diabetes posttransplant. Cell Transplant. 2012;21(12):2797–804.
189. Papas KK, Colton CK, Qipo A, Wu H, Nelson RA, Hering BJ, et al. Prediction of marginal mass required for successful islet transplantation. J Invest Surg. 2010;23(1):28–34.
190. Wang W, Upshaw L, Strong DM, Robertson RP, Reems J. Increased oxygen consumption rates in response to high glucose detected by a novel oxygen biosensor system in non-human primate and human islets. J Endocrinol. 2005;185(3):445–55.
191. Wikstrom JD, Sereda SB, Stiles L, Elorza A, Allister EM, Neilson A, et al. A novel high-throughput assay for islet respiration reveals uncoupling of rodent and human islets. PLoS One. 2012;7(5):e33023.
192. Ricordi C. Quantitative and qualitative standards for islet isolation assessment in humans and large mammals. Pancreas. 1991;6(2):242–4.
193. Bertuzzi F, Garancini P, Socci TC, Nano R, Taglietti MV, Santopinto M, et al. Lessons from in vitro perifusion of pancreatic islets isolated from 80 human pancreases. Cell Transplant. 1999;8(6):709–12.
194. Street CN, Lakey JR, Shapiro AM, Imes S, Rajotte RV, Ryan EA, et al. Islet graft assessment in the Edmonton protocol: implications for predicting long-term clinical outcome. Diabetes. 2004;53(12):3107–14.
195. Ricordi C, Lacy PE, Scharp DW. Automated islet isolation from human pancreas. Diabetes. 1989;38 Suppl 1:140–2.
196. Grant AM, Christie MR, Ashcroft SJ. Insulin release from human pancreatic islets in vitro. Diabetologia. 1980;19(2):114–7.
197. Gerling IC, Kotb M, Fraga D, Sabek O, Gaber AO. No correlation between in vitro and in vivo function of human islets. Transplant Proc. 1998;30(2):587–8.
198. Kayton NS, Poffenberger G, Henske J, Dai C, Thompson C, Aramandla R, et al. Human islet preparations distributed for research exhibit a variety of insulin-secretory profiles. Am J Physiol Endocrinol Metab. 2015;308(7):E592–602.
199. Bertuzzi F, Ricordi C. Prediction of clinical outcome in islet allotransplantation. Diabetes Care. 2007;30(2):410–7.
200. London NJ, Thirdborough SM, Swift SM, Bell PR, James RF. The diabetic "human reconstituted" severe combined immunodeficient (SCID-hu) mouse: a model for isogeneic, allogeneic, and xenogeneic human islet transplantation. Transplant Proc. 1991;23(1 Pt 1):749.
201. Sabek OM, Fraga DW, Minoru O, McClaren JL, Gaber AO. Assessment of human islet viability using various mouse models. Transplant Proc. 2005;37(8):3415–6.
202. Caiazzo R, Gmyr V, Kremer B, Hubert T, Soudan B, Lukowiak B, et al. Quantitative in vivo islet potency assay in normoglycemic nude mice correlates with primary graft function after clinical transplantation. Transplantation. 2008;86(2):360–3.
203. Pepper AR, Gall C, Mazzuca DM, Melling CW, White DJ. Diabetic rats and mice are resistant to porcine and human insulin: flawed experimental models for testing islet xenografts. Xenotransplantation. 2009;16(6):502–10.
204. Merino JF, Nacher V, Raurell M, Biarnes M, Soler J, Montanya E. Optimal insulin treatment in syngeneic islet transplantation. Cell Transplant. 2000;9(1):11–8.
205. Weber DJ. FDA regulation of allogeneic islets as a biological product. Cell Biochem Biophys. 2004;40(3 Suppl):19–22.
206. Bruni A, Gala-Lopez B, Pepper AR, Abualhassan NS, Shapiro AJ. Islet cell transplantation for the treatment of type 1 diabetes: recent advances and future challenges. Diabetes Metab Syndr Obes Targets Ther. 2014;7:211–23.

Index

M. Ramírez-Domínguez (ed.), *Pancreatic Islet Isolation*, Advances in Experimental Medicine and Biology 938, DOI 10.1007/978-3-319-39824-2

FSC
www.fsc.org
MIX
Papier | Fördert gute Waldnutzung
FSC® C083411

Kolibri 360 GmbH
Ferdinand-Jühlke-Straße 7
99095 Erfurt, Deutschland
produktsicherheit@kolibri360.de